Tomorrow's Promise

Book 4: Living with dementia

Dr. Patricia A. Farrell

Book cover photo: Joshua@unsplash.com

Amazon author page for P. A. Farrell: https://tinyurl.com/4zk9pfe6 AND https://tinyurl .com/t8xkp4zx

The Pocket Companion Series contains:

When Your Mind Won't Stop

After the Loss: Finding Your Way Through Grief

You Are Enough: Rebuilding Your Self-Worth

At the Crossroads: Making Decisions When Nothing Feels Clear

When People Hurt: Navigating Difficult Relationships

When You Feel Stuck: Finding Movement in Hard Times

Books by Patricia A. Farrell, Ph.D.

DR. PATRICIA A. FARRELL

Better Together, Book 3: Reasons we heal

The Belief Effect, Book 2: Reasons we heal

Wired for Good, Book 1: Reasons we heal

Get Out! A shrink's guide to using the great outdoors for therapy

When You Can't Pour From an Empty Glass: CBT Skills for Exhausted Caregivers

The Little Book on Learning Big Critical Thinking Skills

The Smart Kid's Survival Guide: Making Good Choices in a Confusing World

How to Be Your Own Therapist

It's Not All in Your Head: Anxiety, Depression, Mood Swings and Multiple Sclerosis

Unfiltered: Beneath the noise of our thoughts lies the true narrative of our minds

Unfiltered Again: A behind-the-scenes look at healthcare, medicine and mental health

A Social Security Disability Psychological Claims Handbook: A simple guide to understanding your SSD claim for psychological impairments and unraveling the maze of decision-making

A Social Security Disability Psychological Claims Guidebook for Children's Benefits

The Disability Accessible US Parks in All 50 States: A Comprehensive Guide

Birding in the US NOW!: A birding guide for individuals with disabilities

Contents

Author's Note

I want to be honest with you about something before we begin.

I've written three books in this series about *what dementia is, how to live with it, and how to care for someone who has it*. Those books are about the present. *This one is about the future*. And writing about the future in medicine requires a particular kind of care, because the temptation to overstate hope is real, and the people reading this book deserve better than that.

So here is my promise to you: everything in this book is grounded in actual science, not wishes. When I say there is reason for hope, I mean it. When I say we're not there yet, I mean that too. The story of dementia research right now is *genuinely exciting*, and I don't need to exaggerate it.

The last decade has produced more progress in dementia research than the four decades before it combined. We now have drugs that can slow Alzheimer's disease, not just manage symptoms. We have blood tests that can detect Alzheimer's changes in the brain before symptoms appear. We have a clearer understanding than ever before of what causes the disease and what might stop it.

We don't have a cure. I want to be clear about that. But we are closer than we have ever been. And the people reading this book today, living with a diagnosis or loving someone who is, are living in the most promising moment in the history of this disease.

This book is for you. It will tell you what's happening in research, what's coming, and what you can do to be part of it.

A Note About Currency of Research

Dementia research moves faster than books can be updated. By the time you read these pages, some specific trial results will have changed, some drugs will have moved through approval, and new discoveries will have been made. Everything is at an accelerated pace thanks to innovations that have been brought about by artificial intelligence. Where it would take months or years to make these major new discoveries, AI has made them within weeks to months. It is startling.

The principles in this book, about how clinical trials work, why certain biological targets matter, what prevention science shows, and how to evaluate research claims, will remain valid even as specific details evolve.

At the end of this book, you'll find a list of the most reliable sources for staying current on new developments. Those sources will do what this book cannot: update in real time.

Read this book for the framework. *Go to those sources for the latest.*

Chapter 1: Why This Is the Most Hopeful Moment in Dementia History

In 1906, a German physician named Alois Alzheimer examined the brain of a patient who had died after years of progressive memory loss, confusion, and behavioral change.

What he saw under the microscope would bear his name for the next century: clumps of abnormal protein deposits between brain cells, and twisted fibers inside them. He had identified, for the first time, the physical signature of a disease that would go on to affect hundreds of millions of people worldwide.

He had also, in a sense, started a clock. The clock of how long it would take to move from identifying the disease to understanding it well enough to slow it down.

That clock ran for 116 years. And then, in 2023, something changed.

A Century of Effort and Setback

For most of the twentieth century, dementia was considered an inevitable part of aging. The term 'senility' captured the attitude of the era: a sad but unavoidable decline that came with old age and could not be meaningfully addressed. Research was sparse. Funding was minimal. The disease was managed, not studied.

That began to change in the 1970s and 1980s, when scientists started taking Alzheimer's disease seriously as a biological problem to be solved rather than a condition to be endured. The discovery that the plaques and tangles Alois Alzheimer had described consisted of specific proteins, amyloid and tau, gave researchers a target. And what followed was several decades of intensive effort to develop drugs that could address those targets.

It did not go well, at first. Between 1998 and 2017, more than 146 clinical trials for Alzheimer's drugs failed. Drugs that worked beautifully in mice didn't work in people. Drugs that cleared amyloid from the brain didn't improve cognitive function. The pharmaceutical industry began pulling back from dementia research because the failure rate was so high. Scientists began questioning the fundamental theories about what was driving the disease.

This period of failure was painful for researchers, devastating for families, and genuinely discouraging. But it was also productive in ways that weren't immediately visible. *Every failed trial taught scientists something about what didn't work and, more importantly, began to reveal why.* The field was learning, even as it was failing.

Why Progress Was So Slow

Several factors made Alzheimer's research uniquely difficult.

First, the disease begins in the brain years, sometimes decades, before any symptoms appear. By the time a person received a diagnosis, the damage in their brain was already extensive. The drugs being tested were arriving, in many cases, too late, trying to put out a fire that had been burning for twenty years.

Second, there were no reliable biological markers that could confirm Alzheimer's disease in a living person without an invasive spinal tap or an expensive specialized brain scan. This made it difficult to enroll the right people in trials and to measure whether drugs were actually hitting their intended target.

Third, the brain is extraordinarily complex and extraordinarily protected. The blood-brain barrier, which is the biological fence that prevents most substances from entering the brain from the bloodstream, makes getting drugs into the brain much harder than getting them into other organs.

Each of these problems has now been meaningfully addressed. Blood tests that can detect Alzheimer's-related proteins were FDA-cleared in 2024, making early biological diagnosis far more accessible. Amyloid PET scans, while expensive, allow researchers to see amyloid in the living brain. And the new generation of anti-amyloid antibody drugs was specifically designed to cross the blood-brain barrier and reach its target.

The Turning Point

The turning point came in the form of two drugs and the large clinical trials that proved they worked.

Lecanemab, tested in a trial called CLARITY AD, was the first drug to show statistically significant slowing of cognitive and functional decline in people with early Alzheimer's disease in a phase 3 clinical trial. The FDA approved it in January 2023 on an accelerated basis, and gave it full approval in July 2023. It was the first Alzheimer's drug in history that could be said to change the course of the disease.

Donanemab, tested in a trial called TRAILBLAZER-ALZ 2, followed. Its results were announced in 2023 and the FDA approved it in July 2024. The trial results showed comparable, and in some analyses slightly greater, slowing of decline.

Neither drug is a cure. Neither stops Alzheimer's disease. Neither reverses any damage that has already been done. But both slow the rate at which the disease progresses in people who receive treatment early enough. And that slowing, modest in absolute terms, is the first proof that the biology of this disease can be meaningfully interrupted by a drug. That is a genuine turning point.

Why Even Modest Progress Matters

When the clinical trial results for lecanemab and donanemab were reported, some commentators questioned whether the magnitude of benefit, roughly a 27 to 35 percent slowing of decline over 18 months, was clinically meaningful. The people who raised that question were applying the wrong standard.

A 27 percent slowing of decline isn't a small thing when it's applied to a disease that lasts a decade or more. Extended over several years, that difference means more time in the early stage of the disease, when independence and quality of life are highest. It means more time for future, better treatments to build on. It means proof of concept, the single most important milestone in drug development, which is that this approach can work.

Think of it this way. The first airplanes flew for 12 seconds and 120 feet. By the standards of what aviation would eventually become, that was a laughable achievement. But it proved the concept. It proved that heavier-than-air flight was possible, and everything that followed built on that proof. The first disease-slowing drugs for Alzheimer's are the Wright Brothers moment of this field.

The Scale of the Current Investment

The level of resources being invested in dementia research right now is unprecedented. The National Institutes of Health funding for Alzheimer's and related dementias has grown from roughly $600 million per year in 2015 to over $3.5 billion per year by 2024. The National Plan to Address Alzheimer's Disease, first enacted in 2011 and updated regularly, sets federal targets for research, care, and public health.

Pharmaceutical companies that retreated from dementia research in the early 2000s because of the high failure rate are back, energized by the proof that the amyloid approach can work and by the enormous market that effective treatment would represent.

Private foundations, including the Alzheimer's Association, the Alzheimer's Drug Discovery Foundation, and the Bill and Melinda

Gates Foundation, have committed hundreds of millions of dollars to research outside the traditional pharmaceutical development pathway.

The result is a research pipeline of unprecedented breadth: hundreds of clinical trials, dozens of drug candidates at different stages of development, and multiple biological targets being investigated simultaneously. *The era of a handful of underfunded researchers working against a wall of skepticism is over.* The era of large-scale, well-funded, coordinated global research is here.

That doesn't mean the answers will come quickly or easily. But it means more people are looking, with better tools, than ever before. And that matters.

The Researchers Who Stayed

It's worth taking a moment to understand what the decades of failed clinical trials actually meant for the people doing the research, because the human dimension of this story is important.

Many researchers spent their entire careers working on Alzheimer's disease during the long period of failure. They entered the field as young scientists, attracted by the complexity of the problem and the importance of solving it. They watched their drug candidates work beautifully in mice and fail in people. They built careers on hypotheses that turned out to be wrong or incomplete. They wrote grant applications year after year for research that kept hitting walls.

Some of them left. The attrition from the dementia research field was significant during the worst years of trial failures. But many stayed. They stayed because they believed the problem was solvable and because the alternative, walking away, was not something they were willing to accept given what was at stake.

The researchers who stayed refined the hypotheses. They figured out that drugs were being tested too late. They developed the blood tests and PET scans needed to detect the disease earlier. They designed trials that enrolled the right people at the right time. And eventually, their persistence produced results that wouldn't have come from the people who left.

This is worth knowing because it's a model for how difficult scientific problems get solved: not by quick breakthroughs, but by sustained commitment in the face of failure. The people alive with Alzheimer's today, and those who will be diagnosed in the years to come, are the beneficiaries of that sustained commitment.

Chapter 1 Summary

Dementia research has a 116-year history, most of which involved extensive failure. Progress was slowed by the fact that the disease begins long before symptoms appear, by the absence of reliable biological markers, and by the difficulty of getting drugs into the brain. All of those barriers have now been meaningfully reduced. The FDA approval of lecanemab in 2023 and donanemab in 2024 represents the first proof that the biology of Alzheimer's disease can be interrupted by a drug. Even modest slowing is significant across the long course of the disease and proves the concept that further improvement is possible. Funding for dementia research is at an all-time high, and the research pipeline has never been more active.

Three Things You Can Do Right Now

1. Read the Alzheimer's Association's annual Alzheimer's Disease Facts and Figures report, which summarizes where research and care stand each year. It's available free at alz.org.

2. Share what you learned in this chapter with someone who still believes nothing can be done about dementia. The story has changed, and people deserve to know it.

3. Visit clinicaltrials.gov and search 'Alzheimer's disease' to get a sense of how many trials are currently active. You don't need to enroll. Just seeing the scale can shift what feels possible.

Chapter 2: The Biology of Dementia, Revisited

To understand where research is going, you have to understand what researchers are targeting.

This chapter revisits the biology of dementia, but at a slightly deeper level than Book One covered. In Book One, we focused on what the brain looks like in dementia and what happens to daily function. Here we go one layer deeper: into the molecular processes that scientists are trying to interrupt, and why those specific processes matter for the treatments being developed right now.

You don't need a science background for this. You need the same thing you need for any complex subject: willingness to follow an explanation that uses real examples and plain language. That's what this chapter offers.

Amyloid: The Protein That Started the Story

Amyloid is a protein that exists in the brains of all healthy people. In its normal form, it appears to play a role in protecting neurons and supporting brain function. The problem arises *when the brain produces an abnormal form of amyloid*, specifically a version called amyloid-beta 42, that tends to clump together instead of being cleared away.

Think of a kitchen drain. Normally, water and small particles flow through and drain away. But if something sticky and accumulating gets into the drain over time, it begins to build up into a clog. In Alzheimer's disease, amyloid-beta 42 is the sticky substance. It accumulates between neurons in clumps called plaques, and those plaques interfere with the communication between brain cells.

This buildup doesn't happen overnight. Research has shown that amyloid plaques *begin accumulating in the brain 15 to 20 years before any symptoms* of Alzheimer's disease appear. By the time someone notices memory problems significant enough to see a doctor, their brain has been accumulating amyloid for roughly two decades.

This timeline has profound implications for treatment. A drug that clears amyloid when symptoms are already significant may be arriving too late, like arriving at a house fire after the walls have already burned through. A drug that intervenes when amyloid is just beginning to accumulate, or that prevents the accumulation from starting, has a much better chance of preserving function. This is why the field has shifted so heavily toward early intervention and prevention.

Tau: The Second Target

Tau is a protein that plays a critical role in the internal structure of neurons. Think of neurons as having an internal highway system that carries nutrients and other essential materials from the cell body to the tips of its branches and back again. Tau protein acts like the cross-ties on a railroad track, keeping the rails of that highway in place and the system running smoothly.

In Alzheimer's disease, tau becomes abnormally chemically modified, a process called hyperphosphorylation. When this happens, tau proteins detach from the internal highway rails and instead stick to each other, *forming twisted clumps called neurofibrillary tangles*. The internal highway system collapses. Neurons can no longer get what they need to survive. They die.

Tau tangles tend to spread through the brain in a predictable pattern, and tracking where tangles have spread gives researchers a way to stage the disease. Brain regions with the most tangles show the most significant cell death. The spreading nature of tau pathology also suggests that if the spread can be stopped, the progression of cell death might be slowed.

This is why tau-targeted therapies are one of the most active areas of current research. Several drugs designed to reduce tau accumulation or its spread are in clinical trials, and some of the most promising anti-amyloid drugs also appear to reduce tau pathology as a secondary effect.

Neuroinflammation: The Third Piece

For decades, the focus of Alzheimer's research was almost entirely on amyloid and tau. More recently, a third biological process has emerged as critically important: neuroinflammation, which is inflammation within the brain itself.

The brain has its own immune cells, called microglia, that function as the brain's cleanup crew. In a healthy brain, microglia clear debris, dead cells, and abnormal proteins. They help maintain the environment that neurons need to function. In Alzheimer's disease, microglia are activated by the presence of amyloid plaques and damaged neurons, and they mount an inflammatory response.

The problem is that *chronic, sustained neuroinflammation causes its own damage*. Inflammatory molecules released by activated microglia can harm nearby neurons. The cleanup crew, working overtime for years, begins to damage the very house it's trying to protect.

Several genes associated with increased Alzheimer's risk are related to immune function, not to amyloid or tau. This has led researchers to believe that the immune system's response to amyloid and tau may be a major driver of neurodegeneration, separate from the plaques and tangles themselves. Drugs targeting neuroinflammation are now a significant part of the research pipeline.

This discovery also helps explain something that has puzzled researchers for years: why some people accumulate significant amyloid in their brains but show little or no cognitive decline. Differences in how the brain's immune system responds to amyloid may be a key factor in determining who develops symptomatic dementia and who doesn't.

Synaptic Dysfunction and Vascular Contributions

Two additional biological processes are important to understand.

Synaptic dysfunction refers to problems with the synapses, *the tiny gaps between neurons where communication happens.* Before neurons die in Alzheimer's disease, their synapses begin to malfunction. The chemical signals that carry information across the synapse become unreliable. This synaptic dysfunction is closely linked to early cognitive symptoms, because it affects how efficiently the brain processes information even before significant cell death has occurred.

Vascular health refers to the health of the blood vessels that supply the brain. The brain uses about 20 percent of the body's blood supply and is extremely sensitive to disruptions in that supply. Damaged or narrowed blood vessels reduce oxygen and nutrients to brain tissue, contributing to cell death and cognitive decline. Vascular factors are involved in most forms of dementia, not just vascular dementia specifically. *High blood pressure, diabetes, high cholesterol, and other cardiovascular risk factors* all affect brain blood vessel health, which is one reason managing these conditions matters so much for brain health.

Understanding these multiple biological contributors is shifting dementia research away from a single target approach toward combination approaches, much like HIV treatment evolved from single drugs to combination therapy that attacks the virus at multiple points simultaneously.

Why the Biology Matters for You

You might be wondering why a chapter about cellular biology belongs in a book aimed at people who may not have any science background. Here's why.

Understanding what researchers are targeting helps you make sense of what you read in the news. When a headline says a new drug failed in a clinical trial, you can now ask: *which target was it aimed at? At what stage of disease was it tested? Was it tested in people who already had extensive brain damage, or in people with early-stage changes?* They're all important questions because they help you understand why something may have failed and where it went wrong in the process of correcting cell death and miscommunication in the brain.

Understanding the biology also helps you understand why lifestyle factors matter so much in prevention. Exercise increases blood flow to the brain and has been shown to reduce amyloid accumulation and neuroinflammation. Sleep is when the brain clears waste products including amyloid through a system called the glymphatic system. High blood pressure damages brain blood vessels. These aren't random health recommendations. They're biologically connected to the specific processes that drive dementia.

And understanding that the disease begins in the brain long before symptoms appear *makes the case for early intervention and prevention* in a way that is hard to ignore. The best time to protect the brain is before damage is visible. That time may be right now.

Biomarkers: Seeing the Disease Before It Shows

A biomarker is a measurable biological indicator of a disease or condition. One of the most significant advances in Alzheimer's research in the last decade has been the development of biomarkers that can detect the biological changes of Alzheimer's disease in living people, often years or even decades before any symptoms appear.

Amyloid PET scans use a radioactive tracer that binds to amyloid plaques in the brain. When the scan is read, areas of high amyloid appear as bright regions on the image. A positive amyloid PET scan confirms that Alzheimer's pathology is present in the brain. This test has been available for over a decade and has transformed clinical research.

Tau PET scans, more recently developed, can image tau tangle accumulation in the brain. Because tau pathology correlates more closely with cognitive symptoms than amyloid does, tau PET provides important information about disease stage and progression.

Blood biomarkers are the newest and most accessible development. Several proteins that reflect Alzheimer's brain pathology can now be measured in a blood sample. Plasma amyloid ratios, plasma phosphorylated tau variants, and glial fibrillary acidic protein (all sounds like a foreign language, I know) are among the markers being studied and, in some cases, approved for clinical use. Blood-based biomarker tests cleared by the FDA in 2024 have made biological confirmation of Alzheimer's disease far more accessible than it was when only PET scans and spinal fluid tests were available.

These biomarkers are changing clinical care in two ways. *First, they allow earlier and more accurate diagnosis*, getting the right people onto appropriate treatments before symptoms become severe. *Second, they allow clinical trials to enroll participants* who are selected based on biological criteria rather than symptom criteria alone, which makes trials more efficient and more likely to find effective treatments.

Biology Beyond Alzheimer's: What Drives Other Dementias

While this book focuses primarily on Alzheimer's disease because it is the most common form of dementia and the one with the most active research, the biology of other major dementia types is also the subject of significant scientific attention.

Lewy body dementia and Parkinson's disease dementia are driven by the *abnormal accumulation of a protein called alpha-synuclein in neurons*. Alpha-synuclein, like amyloid, is a normal protein in the healthy brain. When it misfolds and aggregates, it forms the Lewy bodies that give this form of dementia its name. Research into alpha-synuclein follows the same logic as amyloid research: if the abnormal aggregation can be prevented or cleared, the disease may be slowed.

Frontotemporal dementia involves different proteins depending on the subtype. In some forms, tau protein is again the culprit, though in a different distribution than in Alzheimer's disease. In others, a protein called TDP-43 forms abnormal aggregates. And in the forms caused by specific genetic mutations, the targets are those mutated genes and their downstream effects. Frontotemporal dementia is *much rarer than Alzheimer's disease* but disproportionately affects people in their 40s and 50s, making it a research priority for different reasons.

Vascular dementia, as its name suggests, is driven primarily by blood vessel disease in the brain rather than protein aggregation. This means its biology and its preventive strategies are substantially different from Alzheimer's. *Controlling blood pressure, treating diabetes, stopping smoking, and managing other cardiovascular risk factors* are the primary biological interventions available for reducing vascular dementia risk.

Understanding the specific biology of the type of dementia your family has been affected by helps you ask the right questions about relevant research and applicable prevention strategies.

Chapter 2 Summary

Alzheimer's disease involves at least *three major biological processes*: **amyloid plaque** accumulation between neurons, **tau tangle formation** inside neurons, and **neuroinflammation** driven by the brain's own immune cells. Amyloid begins accumulating 15 to 20 years before symptoms appear, which is why early intervention and prevention are the focus of current research. Tau tangles spread through the brain in a predictable pattern and are the target of several drugs in development. Neuroinflammation may be a major independent driver of neurodegeneration, explaining why some people with amyloid plaques never develop symptoms. Synaptic dysfunction and vascular health also contribute to the disease. Understanding these processes helps make sense of research news, explains why lifestyle factors matter, and makes the case for acting early.

Three Things You Can Do Right Now

1. The next time you see a news headline about a dementia drug trial, ask three questions: which biological target was the drug aimed at? At what stage of disease were participants enrolled? And what was the primary outcome that was measured?

2. Look at the cardiovascular risk factors in your own life: blood pressure, blood sugar, cholesterol, smoking, weight. These are directly connected to the brain biology described in this chapter. Managing them is brain protection.

3. Learn about the glymphatic system and sleep. A simple search of 'glymphatic system sleep dementia' will bring up accessible explana-

tions of how sleep clears amyloid from the brain. This is biology you can act on tonight.

Chapter 3: The Drugs That Are Changing Everything

For a very long time, the question in Alzheimer's drug development was whether it was possible to make a drug that changed the course of the disease. Not just managed symptoms. Not just made people feel slightly better for a period of time. But actually slowed the progression of the underlying disease.

The answer, as of 2023, is yes. It is possible. Two drugs have proven it. This chapter explains how they work, who they're for, what they can and can't do, and what comes next.

How Anti-Amyloid Antibody Drugs Work

Both *lecanemab and donanemab* belong to a class of drugs called *anti-amyloid monoclonal antibodies*. Understanding what that means helps make sense of how they work.

An antibody is a protein that the immune system uses to identify and neutralize foreign invaders. Normally, antibodies are made by the immune system in response to infections. Scientists have learned how to engineer antibodies in the laboratory that are specifically designed to recognize and bind to a particular target.

Lecanemab and donanemab are antibodies *engineered to bind to amyloid in the brain*. When these drugs are given by intravenous infusion, the antibodies travel through the bloodstream, cross the blood-brain barrier, and bind to amyloid. Once bound, *they flag the amyloid for removal by the brain's immune cells.* Over months of treatment, this process clears a significant amount of amyloid from the brain.

The two drugs target slightly different forms of amyloid. Lecanemab targets a form called protofibrils, which are smaller, more soluble amyloid aggregates. Donanemab targets amyloid plaques that contain a particular form of modified amyloid called pyroglutamate amyloid, which tends to be present in more established plaques.

Both drugs are given as intravenous infusions, typically every two weeks for lecanemab and every four weeks for donanemab. Treatment requires ongoing infusions, regular monitoring with brain MRI, and in most cases, amyloid PET scanning or blood biomarker testing to confirm amyloid pathology before starting treatment.

What the Clinical Trials Showed

The CLARITY AD trial for lecanemab enrolled 1,795 people with early Alzheimer's disease, meaning *people with mild cognitive impairment or mild dementia* due to Alzheimer's. Participants were treated for 18 months. The trial found that lecanemab slowed the rate of cognitive and functional decline by approximately 27 percent compared to placebo.

The TRAILBLAZER-ALZ 2 trial for donanemab enrolled 1,736 *people with early Alzheimer's.* The trial found that donanemab slowed decline by approximately 35 percent overall, with even larger effects in people who had lower levels of tau tangles at the start of the trial, suggesting that *the drug works better when the disease is caught earlier.*

In both trials, the treatment also *significantly reduced amyloid in the brain,* confirming that the drugs were doing what they were designed to do. The relationship between amyloid reduction and clinical benefit was not perfect, but it was present. Clearing amyloid does appear to slow the clinical progression of the disease.

What do these numbers mean in daily life? In both trials, the difference between treated and untreated groups on individual cognitive or functional tasks was modest and might not be noticeable to a casual observer. But across multiple outcomes and over time, the *treated groups clearly and consistently did better.* The treated groups had more time in the early stage of the disease, more time with preserved independence.

ARIA: The Side Effect That Requires Monitoring

The most important side effect of anti-amyloid antibody drugs is a set of changes visible on brain MRI called ARIA, which stands for *amyloid-related imaging abnormalities.* ARIA appears to be related to the process of amyloid being cleared from blood vessel walls in the brain, which *can cause small amounts of fluid accumulation or tiny bleeds.*

ARIA comes in two forms. ARIA-E involves edema, which is fluid accumulation in or around brain tissue. ARIA-H involves hemosiderin deposits, which are evidence of small bleeds. Both are detected by MRI, not necessarily by symptoms.

In both lecanemab and donanemab trials, ARIA was common. In the lecanemab trial, ARIA occurred in approximately 21 percent of treated participants. In the donanemab trial, the rate was similar. Most cases of ARIA were detected on scheduled MRI scans and *caused no symptoms.* Approximately 3 percent of participants had symptomatic ARIA, with *symptoms including headache, dizziness, confusion, or nausea*. Serious cases, including large bleeds, were *rare but did occur.*

Because *ARIA risk is higher in people who carry the APOE4 gene variant*, genetic testing for APOE4 is typically part of the evaluation before starting treatment. APOE4 carriers, particularly those with two copies of the gene variant, face higher ARIA risk and require more careful monitoring.

Regular brain MRI monitoring is required for everyone taking these drugs, typically every few months in the first year of treatment. This is a real logistical and financial burden, and it contributes to the complexity of accessing these treatments.

Who Is and Who Is Not a Candidate

Anti-amyloid therapies are not for everyone with dementia. They're currently *indicated for a specific population: people with early-stage*

Alzheimer's disease who still have mild symptoms and who have confirmed amyloid pathology in their brain.

Early-stage Alzheimer's means people with mild cognitive impairment due to Alzheimer's or mild Alzheimer's dementia. *People with moderate or severe dementia were not included in the trials*, and the drugs have not been shown to help people in those stages.

Amyloid confirmation means a positive result on either an amyloid PET scan or a validated blood biomarker test. Not everyone who has memory problems has Alzheimer's disease, and treating someone without confirmed Alzheimer's pathology would expose them to ARIA risk without potential benefit.

Several medical conditions may make these drugs inappropriate, including certain blood-clotting disorders, recent strokes or bleeding in the brain, and the need for blood-thinning medications in high doses. A careful medical evaluation is required before starting treatment.

Cost and insurance coverage are significant barriers. These drugs cost tens of thousands of dollars per year, and Medicare coverage, while improved, comes with conditions and cost-sharing that make access difficult for many patients. The coverage landscape continues to evolve, and the most current information on coverage should be sought from the physician and from the Alzheimer's Association.

The Next Wave: What's in the Pipeline

Lecanemab and donanemab are the beginning of the anti-amyloid story, not the whole story. The pipeline includes several drugs targeting amyloid in different ways, drugs targeting tau, drugs targeting neuroinflammation, and combination approaches that address multiple pathways simultaneously.

Anti-tau drugs are perhaps the most closely watched part of the pipeline after the anti-amyloid antibodies. Several antibodies designed to target tau and prevent its spread through the brain are in Phase 2 and Phase 3 trials. Because tau pathology more closely tracks with cognitive symptoms than amyloid pathology does, *successfully targeting tau may produce more robust clinical benefit.*

Neuroinflammation-targeted drugs, including drugs that modify the activity of microglia, are in active development. These approaches may prove to be important additions to anti-amyloid treatment, addressing a biological process that anti-amyloid drugs don't directly target.

Combination therapy trials, which test two or more drugs together, are beginning, drawing on the lesson from HIV treatment that attacking a complex disease from multiple angles simultaneously is more effective than any single approach.

The field is moving faster than it has at any point in its history. That is not promotional language. It is a description of what's happening in research laboratories and clinical trial centers right now.

Drugs for Other Dementia Types

Much of the most visible drug development has focused on Alzheimer's disease, which accounts for the majority of dementia cases. But research for other types of dementia is also advancing, and people living with Lewy body dementia, frontotemporal dementia, and vascular dementia have reasons for cautious optimism.

Lewy body dementia has been historically understudied relative to its prevalence, in part because it was difficult to diagnose and often misclassified. Research into alpha-synuclein, the protein that forms

Lewy bodies, has accelerated significantly. Several drugs targeting alpha-synuclein are in clinical trials for Parkinson's disease dementia and Lewy body dementia. The development of biomarkers for alpha-synuclein pathology, including a spinal fluid test and skin biopsy test now in clinical use, has *transformed the ability to diagnose Lewy body conditions accurately.*

Frontotemporal dementia has a different target: the tau and TDP-43 proteins that drive different subtypes of the disease. Several trials are underway for FTD subtypes, including antisense oligonucleotide drugs targeting the GRN and C9orf72 genes that cause some familial forms of FTD. These trials are at earlier stages than the Alzheimer's work, but they reflect a genuine investment in a condition that has been largely without treatment options.

Vascular dementia research is closely tied to cardiovascular medicine. The most powerful intervention known for preventing vascular dementia is *aggressive management of the vascular risk factors* that cause it: blood pressure, diabetes, cholesterol, and smoking. Research is exploring whether certain blood pressure medications, particularly those that cross the blood-brain barrier, offer additional brain protection beyond blood pressure reduction alone.

The common thread across all of these is that *the era of dementia as an untreatable condition is ending*. Not at the same pace for every type, and not without significant work still ahead. But the direction is unambiguous.

Chapter 3 Summary

Lecanemab and donanemab are anti-amyloid monoclonal antibodies that work by binding to amyloid in the brain and flagging it for removal by the brain's immune cells. Clinical trials showed that both drugs slowed cognitive and functional decline by 27 to 35 percent over 18 months in people with early Alzheimer's disease. The most significant side effect is ARIA, changes visible on MRI that are usually asymptomatic but require regular monitoring. These drugs are not for everyone: they are indicated for early-stage Alzheimer's patients with confirmed amyloid pathology, and several medical conditions and APOE4 carrier status affect eligibility and risk. Cost and insurance coverage remain significant barriers. The research pipeline includes anti-tau drugs, neuroinflammation-targeted therapies, and combination approaches, with more than one large trial expected to report results in the coming years.

Three Things You Can Do Right Now

1. Ask the neurologist at the next appointment whether anti-amyloid therapy might be appropriate to evaluate. If the answer is no, ask specifically why, and what criteria would need to change.

2. Look up the current Medicare coverage policy for lecanemab and donanemab on the Medicare website or through the Alzheimer's Association, as coverage guidelines change frequently.

3. Search 'Alzheimer's anti-tau clinical trials' on clinicaltrials .gov to see what's currently enrolling. Even if you're not a candidate today, knowing what's active helps you track the field.

Chapter 4: Clinical Trials: What They Are and How to Join One

Every treatment that has ever been approved by the FDA got there through clinical trials.

Every drug that has helped any patient, anywhere, was first tested on people who volunteered to try something that was not yet proven, because they believed that helping to answer the question was worth something. The drugs that are now slowing Alzheimer's disease exist because thousands of people enrolled in clinical trials, sat through infusions and MRI scans, kept symptom diaries, and came back to the clinic month after month.

This chapter explains how clinical trials work, how to find out whether you or someone you love might be a candidate for one, and what participating involves.

The Four Phases of Drug Development

Before a drug can be tested in a clinical trial with patients, it goes through years of laboratory research and animal studies. If those results are promising, the drug moves into human testing, which happens in four phases.

Phase 1 trials focus on safety. A small number of healthy volunteers or people with the disease, typically between 20 and 100 participants, receive the drug at various doses to determine what dose is safe, how the body processes the drug, and what side effects occur. *Phase 1 is about figuring out whether the drug can be given to humans without causing serious harm.*

Phase 2 trials focus on preliminary effectiveness and dose finding. A larger group, typically a few hundred participants, receives the drug to see whether there are early signs it might work and to identify the dose that produces the best balance of benefit and safety. Phase 2 trials often generate the hypothesis that Phase 3 will test.

Phase 3 trials are the *large, definitive trials that produce the evidence needed for FDA approval.* They typically enroll hundreds to thousands of participants, are randomized and controlled, and run for months to years. These are the trials that produced the results for lecanemab and donanemab.

Phase 4 trials *happen after approval*, when a drug is on the market. They study the drug in the real world, in larger and more diverse populations, over longer time periods, often identifying side effects or benefits that weren't visible in the controlled trial environment.

What Being in a Trial Actually Means

People have many misconceptions about what participating in a clinical trial involves. The most important one to address is the *fear of being used as a guinea pig.*

Clinical trials are conducted under *strict ethical and regulatory oversight.* Every trial must be approved by an *Institutional Review Board*, which is an independent ethics committee that reviews the trial design to protect participants' rights and safety. Every participant must go through an *informed consent process* in which they are told, in plain language, what the trial involves, what the *potential risks and benefits* are, what alternatives exist, and that *they can withdraw at any time for any reason without penalty.*

Participants are *never charged for the treatment they receive in a trial.* The trial sponsor, typically a pharmaceutical company or government agency, covers the cost of the experimental treatment, the study-related tests, and *often the travel and logistical costs of participating*. Participants may also receive compensation for their time.

In a randomized controlled trial, which is the most common design for large Phase 3 trials, *participants are randomly assigned to either the treatment group or the control group.* The *control* group typically *receives a placebo*, which looks identical to the active drug but *contains no active ingredient*. In most trials, neither the participant nor the research team knows who is in which group until the trial ends. This is called a "double-blind experiment."

Being in the placebo group is not the same as being harmed. Standard of care, meaning the best available existing treatment, continues for all participants. The question being asked *is whether the new treatment is better than standard of care*, not whether the new treatment is better than nothing.

Finding Trials: ClinicalTrials.gov and TrialMatch

Two resources are essential for finding Alzheimer's-related clinical trials.

ClinicalTrials.gov is a federal database of every clinical trial registered in the United States. It includes trials for all conditions and diseases. You can search by condition, location, age, disease stage, and several other criteria. A search for 'Alzheimer's disease' will return hundreds of active trials. The results pages for each trial include a summary of *what's being studied, who is eligible to participate, where the trial is located, and how to contact the research team.*

The Alzheimer's Association's TrialMatch program, available at *trialmatch.alz.org*, is a user-friendly tool specifically designed to match people with Alzheimer's or those at risk with relevant trials. You create a profile by answering questions about your diagnosis, current symptoms, age, and location, and TrialMatch identifies trials that match your profile and provides the contact information for research teams.

A step-by-step approach to searching TrialMatch: go to trialmatch.alz.org. Click 'Find a Trial.' Answer the questions in the profile builder about the diagnosis stage, age, and location. Review the matched trials list. For any trial that looks relevant, click through to see the eligibility criteria in detail. Contact the research team listed for any trial you want to learn more about. The research team can tell you whether you're likely to qualify and what the next steps would be.

Who Is Eligible: Common Criteria

Every trial has specific *eligibility* criteria (both inclusion and exclusion) that define who can and cannot participate. These criteria exist to ensure participant safety and to make sure the trial tests the right people for the drug being studied.

Common inclusion criteria for Alzheimer's trials include: a diagnosis of mild cognitive impairment due to Alzheimer's or mild Alzheimer's dementia; confirmation of amyloid pathology by PET scan or blood test; age range, often between 50 and 90 years; a study partner, usually a family member or caregiver, who can accompany the participant to visits and provide observations about function; and cognitive test scores within a specific range.

Common exclusion criteria include: certain medical conditions that would make the drug unsafe, such as bleeding disorders; recent brain bleeds or strokes; certain psychiatric conditions; other serious medical illnesses; participation in another drug trial within a certain time period (usually called "wash out"); and, for trials specifically targeting early disease, cognitive decline that has progressed too far.

It's worth knowing that most people who inquire about a trial don't qualify. The eligibility criteria are often narrow, and matching all of them is genuinely difficult. This shouldn't be discouraging: it's a reflection of scientific rigor, not a rejection. Contacting multiple trial sites and trying multiple trials is appropriate.

Prevention Trials: If You're Not Diagnosed

Some of the most important clinical trials underway aren't for people who have been diagnosed with dementia. They're for people who have risk factors for Alzheimer's disease but no symptoms yet.

Prevention trials are testing whether interventions, whether drugs, lifestyle changes, or combinations of both, can prevent Alzheimer's disease from developing in people who are at elevated biological or genetic risk. These trials are critical to answering the question of whether early intervention can change the long-term trajectory of the disease.

Several prevention trials are targeting people who carry the APOE4 gene variant, which significantly increases Alzheimer's risk. Others are targeting people with confirmed amyloid accumulation in the brain detected by blood test or PET scan, before any cognitive symptoms have appeared. These are people who might be 10 to 15 years from any symptoms, but who have biological evidence that the disease process has begun.

If you don't have a diagnosis but you have a family history of Alzheimer's or other risk factors, you may be eligible for a prevention trial. TrialMatch includes prevention trials in its search results. This is one of the most impactful things a person at risk can do: contribute to the research that may protect the next generation, while potentially benefiting from interventions that the research uncovers.

The Day-to-Day Reality of Trial Participation

For people considering a clinical trial, it can be helpful to understand what participation actually looks like on a practical level, beyond the official description of procedures and visits.

For a typical Phase 3 Alzheimer's drug trial, participants usually begin with a screening period that involves one or more clinic visits for cognitive testing, blood tests, and often a brain scan to confirm eligibility. This screening process can take several weeks and may involve multiple trips to the trial site. Approximately half or more of people who apply for trials are screened out at some point during this process because they don't meet the eligibility criteria.

Once enrolled, participants typically have regular clinic visits, often monthly or every few months, for the duration of the trial. These visits involve cognitive testing, blood draws, physical examination, and questions about symptoms and side effects. The time commitment is real, often several hours per visit, plus travel time.

For trials involving intravenous drug administration, like the anti-amyloid antibody trials, infusion visits are added to the schedule. Each infusion typically takes one to several hours, plus observation time afterward. For a person receiving infusions every two weeks, this represents a significant ongoing time commitment.

Study partners, usually family members or close friends, are required for most Alzheimer's trials and play an important role. They provide observations about the participant's daily function, accompany them to visits, and are interviewed separately about any changes they've noticed. For caregivers who are already supporting someone with early dementia, this role fits naturally into what they're already doing.

Most participants in clinical trials describe the experience as more manageable than they expected, partly because trial staff provide extensive support and guidance. The relationship between participants and trial staff is often described as one of the most positive aspects of participation: a group of people who are genuinely invested in your well-being and who are paying very close attention to how you're doing.

What Happens After a Trial Ends

One question many people considering clinical trial participation ask is what happens to them when the trial ends. This is a legitimate concern and deserves a clear answer.

When a trial concludes, participants transition back to standard care with their own medical team. The trial sponsor typically provides a summary of results to participants, which explains what the trial found and what their participation contributed. *Some trials offer open-label extension phases,* where participants who wish to continue receiving the experimental treatment can do so while further safety data is gathered.

For people who were in the placebo group, learning what treatment they missed out on can be difficult emotionally. But participation in a placebo group is not wasted. Without the placebo group, there is no way to know whether a treatment actually works. Placebo participants are the reason that the trial produced evidence, and that evidence either validates a treatment or prevents a useless or harmful one from reaching the market.

Some clinical trials have compassionate use programs that allow participants to access a promising treatment before it's approved, if the results are strong enough to suggest benefit and the safety profile is acceptable. The FDA's expanded access program provides a pathway for this.

And all participants, including those in placebo groups, contribute their data to the research record. That data continues to be analyzed for years after the trial ends, informing future trial design and contributing to the growing understanding of the disease. Participation is a gift to the research, regardless of which group you're assigned to.

Chapter 4 Summary

Every approved treatment came through clinical trials, and every clinical trial depends on people who volunteer to participate. Drug development proceeds through four phases: *safety testing, preliminary effectiveness, large definitive trials, and post-approval monitoring*. Participants are protected by ethical oversight, are never charged for the experimental treatment, give informed consent, and can withdraw at any time. ClinicalTrials.gov and the Alzheimer's Association's TrialMatch program are the primary tools for finding relevant trials. Eligibility criteria are specific and often narrow: most inquiries don't result in enrollment, but contacting multiple sites and trials is appropriate. Prevention trials for people at elevated risk who have no symptoms yet are critically important and may be open to people with a family history or known risk factors.

Three Things You Can Do Right Now

1. Visit trialmatch.alz.org this week and create a profile. Even if you don't enroll in anything, knowing what's available and whether you might be eligible is valuable information.

2. At the next neurology appointment, ask specifically whether any clinical trials currently enrolling would be appropriate to consider. Research-oriented neurology centers often know

about trials that aren't widely advertised.

3. If you don't have a diagnosis but have a family history of Alzheimer's, look specifically at prevention trials on Clinica lTrials.gov. Search 'Alzheimer's prevention' and filter by your location.

Chapter 5: The Genetics of Dementia and What It Means for Families

No topic in dementia raises more anxiety in families than genetics.

When a parent or grandparent is diagnosed with Alzheimer's disease, the question that follows is almost immediate: *does this mean I'll get it too?* If you're a middle-aged adult watching a parent decline, that question can feel both urgent and terrifying.

This chapter addresses that question honestly, which means addressing it more carefully than it's often handled in online searches and general conversations. The genetics of dementia are real and they matter. They are also more complicated, and in most cases more reassuring, than the anxiety suggests.

Two Kinds of Genetic Risk

The genetics of Alzheimer's disease involve two fundamentally different categories that are often confused with each other: *risk genes and deterministic genes.*

Risk genes increase the likelihood of developing Alzheimer's disease, but don't guarantee it. Having a risk gene means your odds are higher, not that the outcome is certain. The most significant risk gene is APOE4, and it deserves its own section below.

Deterministic genes cause Alzheimer's disease in everyone who inherits them. These genes, which include mutations in three specific genes called APP, PSEN1, and PSEN2, produce a form of Alzheimer's called *familial Alzheimer's disease* that typically begins in a person's 40s or 50s rather than in their 70s or 80s. This form of Alzheimer's is *rare, accounting for less than 1 percent of all Alzheimer's disease.* But it follows an inheritance pattern: a child of someone with a deterministic gene mutation has a 50 percent chance of inheriting it, and everyone who inherits this genetic combination will develop the disease.

The distinction between these two categories is critical because they carry very different implications for families. Most people with a family history of Alzheimer's disease, including many whose parent or grandparent had the disease, are *dealing with risk gene considerations, not deterministic gene considerations.*

APOE4: The Most Important Risk Gene

The apolipoprotein E gene, known as APOE, comes in three common versions called alleles: APOE2, APOE3, and APOE4. Because everyone inherits one copy from each parent, a person can have any of these combinations: 2/2, 2/3, 2/4, 3/3, 3/4, or 4/4.

APOE4 is the version associated with increased Alzheimer's risk. Having one copy of APOE4, which is called being heterozygous for APOE4, increases risk by roughly two to three times compared to someone with two copies of APOE3. Having two copies of APOE4, called being homozygous, increases risk by roughly eight to twelve times compared to the APOE3/3 baseline.

These sound like large numbers, and in relative terms they are. But it's important to understand what they mean in absolute terms. *Most APOE4 carriers never develop Alzheimer's disease.* Approximately 30 percent of people with one copy of APOE4 develop Alzheimer's disease by age 85, compared to roughly 10 percent of people without APOE4. *That means 70 percent of people with one APOE4 copy do not develop Alzheimer's by that age.*

APOE4 also appears to affect the age of onset and the rate of amyloid accumulation. People with APOE4 tend to accumulate amyloid earlier and at faster rates. This is why APOE4 status is a factor in the risk and monitoring protocols for anti-amyloid drug therapy.

APOE2, by contrast, appears to be protective. People with APOE2 have a lower than average risk of developing Alzheimer's disease. Research is actively *exploring why APOE2 is protective as a possible route to developing protective interventions.*

What Most People with a Family History Need to Know

The most important thing for most people with a family history of Alzheimer's disease to understand is this: having a parent or sibling with Alzheimer's increases your risk, but *most people with an affected parent will not develop Alzheimer's disease.*

Having one first-degree relative, meaning a parent or sibling, with Alzheimer's roughly doubles the risk compared to someone with no family history. Having multiple first-degree relatives with Alzheimer's increases the risk further. But even with a strong family history, the majority of people will not develop the disease.

Family history is also influenced by factors other than genetics. Families share environments, lifestyles, and health behaviors as well as genes. A family in which multiple members have cardiovascular disease, sedentary lifestyles, or poor sleep habits may have elevated Alzheimer's rates partly because of those shared lifestyle factors, not solely because of genetics.

The most practical response to a family history of Alzheimer's is not genetic testing, at least not as a first step. It is *aggressive management of the modifiable risk factors* covered in Chapter Six: blood pressure, exercise, sleep, social engagement, and hearing health. These are factors that are within your control, regardless of your genetic profile.

Genetic Testing: When It Makes Sense

Genetic testing for Alzheimer's risk is available but requires careful consideration of what you're looking for and what you'll do with the information.

Testing for APOE4 through a clinical genetic counselor or through direct-to-consumer services like 23andMe is available. Before testing, it's worth asking: what would I do differently if I knew my APOE4 status? If the answer is that you'd be more motivated to exercise, manage blood pressure, and improve sleep regardless, then testing may provide useful motivation. If the answer is that knowing you're APOE4 positive would cause significant anxiety without changing your behavior, the value is less clear.

Testing for the rare deterministic gene mutations is *typically considered only in people with a strong family history of early-onset Alzheimer's*, meaning disease onset before age 65 in multiple family members. This testing should always be done in conjunction with genetic counseling, because the implications are profound and require appropriate preparation and support.

The REVEAL study, which is one of the most important studies on genetic risk disclosure, examined how people responded to learning their APOE4 status. The study found that most people tolerated the information reasonably well, that disclosure did not cause lasting psychological harm in most participants, and that some people found it useful for planning. But a minority of participants experienced significant anxiety. Genetic counseling before and after testing reduces those adverse responses.

Direct-to-consumer genetic tests, like those from 23andMe or AncestryDNA, can report APOE status, but *they may not test all relevant variants, may provide results without adequate genetic counseling support, and may not be as accurate as clinical laboratory testing.* They are a starting point, not a definitive answer.

What to Do with Genetic Information

Whether or not you pursue genetic testing, there are specific things worth doing with what you know about your family history.

1. Document the family history carefully. Write down which relatives were diagnosed with Alzheimer's or another dementia, at what age they were diagnosed, and how their disease progressed. This information is useful to your physician and relevant if you ever pursue genetic counseling or clinical trial participation.

2. Discuss your family history explicitly with your primary care physician. A family history of Alzheimer's should prompt earlier and more frequent cognitive screening and more aggressive management of cardiovascular risk factors.

3. Consider whether you meet criteria for a prevention trial. People with a family history, particularly with known genetic risk factors, are often the target population for prevention trials that are testing whether interventions can delay or prevent the disease.

4. Keep the genetics in perspective. *Family history is one risk factor* among several, and unlike most of the others, it's one you can't change. Focus your energy on the ones you can: blood pressure, exercise, sleep, social connection, and the other lifestyle factors in the next chapter.

Families with Early-Onset Alzheimer's

For families affected by the rare deterministic gene mutations that cause early-onset familial Alzheimer's disease, the experience is distinct from what most families with Alzheimer's face. These families know, with a degree of certainty that is rare in medicine, that the disease will come if the gene is inherited. And they know it will come early, often in the 40s or 50s, during the years when children are still at home and careers are at their peak.

The *Dominantly Inherited Alzheimer Network, known as DIAN*, is a research program specifically designed for families with these mutations. DIAN has studied several hundred families worldwide with deterministic mutations in APP, PSEN1, and PSEN2, tracking them over time to understand the earliest biological changes that precede symptoms. This research has produced some of the most important insights into the timeline and biology of Alzheimer's disease available anywhere.

DIAN also runs clinical trials specifically for people with these mutations, testing whether drugs given very early, before symptoms begin, can change the course of the disease. These trials represent a unique opportunity for families who face a virtually certain diagnosis: the chance to potentially alter what has always seemed inevitable.

For families living with the knowledge of a deterministic mutation, *genetic counseling is essential,* not just before testing but as an ongoing support. The psychological burden of this knowledge is significant, and the decisions it generates, whether to test, whether to share results with children, whether to participate in research, are deeply personal and deserve expert guidance.

If your family has been affected by early-onset Alzheimer's, ask a neurologist or genetic counselor specifically about DIAN and about whether research programs designed for your situation might be relevant. These families have a particular and powerful role to play in the research that will eventually change outcomes for everyone.

Chapter 5 Summary

Alzheimer's genetics involves two fundamentally different categories: *risk genes like APOE4*, which increase the probability of developing the disease, and *deterministic genes like APP, PSEN1, and PSEN2*, which cause early-onset familial Alzheimer's in everyone who inherits them but are very rare. APOE4, the most significant risk gene, increases risk two to twelve times depending on copy number, but most APOE4 carriers do not develop Alzheimer's disease. Having one first-degree relative with Alzheimer's roughly doubles risk, but the majority of people with an affected parent will not develop the disease. Genetic testing for APOE4 can be informative but should be accompanied by genetic counseling. The most practical response to genetic risk is aggressive management of the modifiable risk factors in Chapter Six.

Three Things You Can Do Right Now

1. Document your family health history this week. Write down which relatives had Alzheimer's or dementia, at what age they were diagnosed, and what you know about how their disease progressed. Share this document with your physician.

2. If you're considering genetic testing for APOE4, ask your doctor for a referral to a genetic counselor first. Talking through the implications before testing prepares you for whatever the result is.

3. Remind yourself that family history is one risk factor you cannot change. Redirect your energy to the ones you can

change, starting with Chapter Six.

Chapter 6: Prevention: What Science Actually Says

The most common question in dementia medicine is *whether the disease can be prevented.*

The honest answer, as of today, is: *not completely, and not with certainty*. We can't guarantee that anyone who does everything right will never develop dementia. The disease is too complex, and the role of factors outside human control, including genetics and simple biological aging, is too significant for that.

But the evidence for risk reduction is genuinely strong. A landmark analysis by the Lancet Commission on Dementia Prevention, Intervention, and Care identified 12 modifiable risk factors that together account for approximately 40 percent of dementia cases worldwide. That means roughly 40 percent of dementia, by current estimates, is potentially preventable by addressing factors within human control.

Forty percent is not a small number. It is an enormous opportunity. And it starts earlier in life than most people think.

The Lancet Commission's 12 Risk Factors

The 2020 Lancet Commission report, updated in 2024, identified *12 modifiable risk factors for dementia across the lifespan*. Together, addressing these factors could reduce the global burden of dementia by approximately 40 percent.

The risk factors are:

1. lower education in early life
2. hearing loss in midlife
3. high blood pressure in midlife
4. obesity in midlife
5. excessive alcohol consumption
6. smoking
7. depression
8. social isolation
9. physical inactivity
10. diabetes
11. air pollution
12. traumatic brain injury

Two things are immediately striking about this list. *First, most of these factors are relevant long before dementia symptoms appear,* in midlife or even earlier. High blood pressure in your 40s and 50s is a more significant dementia risk factor than high blood pressure in your 70s. Education level in childhood affects dementia risk decades later. The brain is being shaped by these factors across an entire lifetime, not just in old age.

Second, several of these factors are well-established cardiovascular risk factors: hypertension, obesity, diabetes, and smoking. The conditions that damage your heart also damage your brain. Managing cardiovascular health is brain health.

The Strongest Modifiable Risk Factors

Among the 12 Lancet Commission factors, *blood pressure in midlife has some of the strongest evidence.* The SPRINT MIND trial found that intensive blood pressure control, bringing systolic blood pressure below 120 mm Hg, significantly reduced the incidence of mild cognitive impairment compared to standard blood pressure targets. The mechanisms are clear: high blood pressure damages blood vessels throughout the body, including the small vessels in the brain, causing cumulative injury over decades.

Hearing loss is the single largest modifiable risk factor for dementia in midlife, accounting for approximately 8 percent of dementia cases. The mechanisms may include the cognitive load of effortful listening that strains brain resources, social withdrawal from difficulty communicating, or direct effects of auditory deprivation on brain structure. Treating hearing loss with hearing aids reduces cognitive load and may protect against this risk. If you or someone you love has hearing loss and has resisted hearing aids, this evidence provides a compelling reason to reconsider.

One of the most compelling reasons people give for not wanting to wear hearing aids is *the impact they have on self-esteem*. There are new designs called "intercanal" hearing aids that are almost invisible and can be worn without being seen. They can be adjusted via a cell phone or an internet connection, and they can be bought over the counter. The costs of these units are dramatically different from what audiologists used to charge. Now hearing aids can be bought for a few hundred dollars rather than several thousand, and they are rechargeable.

Physical inactivity is among the most actionable risk factors. The evidence that aerobic exercise reduces dementia risk is substantial and consistent across dozens of studies. Exercise increases blood flow to the brain, reduces inflammation, promotes the release of brain growth factors, and has been shown in randomized trials to increase the volume of the hippocampus, the brain's memory center. The minimum effective dose appears to be 150 minutes of moderate-intensity aerobic activity per week, equivalent to a brisk 30-minute walk five days a week.

Social isolation is a risk factor that has received increased attention since the COVID-19 pandemic, which created a natural experiment in forced isolation for populations worldwide. Research consistently shows that people with strong social connections have lower rates of dementia than those who are socially isolated, independent of other factors. The mechanisms involve both the cognitive demands of social interaction and the buffer that social connection provides against depression and chronic stress.

Sleep and the Glymphatic System

Sleep deserves its own section in any discussion of dementia prevention, because *the discovery of the glymphatic system* has dramatically changed how scientists think about the relationship between sleep and brain health.

The glymphatic system is the brain's waste clearance system. It's the garbage truck that comes regularly while we sleep and removes metabolic waste. During sleep, particularly during deep sleep, the *spaces between brain cells expand and cerebrospinal fluid is pumped through the brain at an increased rate*, flushing out waste products. One of the waste products cleared by the glymphatic system is *amyloid-beta*, the protein that accumulates into plaques in Alzheimer's disease.

This discovery, made by researcher Maiken Nedergaard and colleagues, means that sleep is not just rest for the brain. *It is active maintenance.* A brain that consistently gets inadequate sleep is a brain that consistently fails to clear amyloid. Over decades, that shortfall may contribute to the accumulation of amyloid that drives Alzheimer's disease.

Studies have found that people who regularly sleep *fewer than six hours per night have higher rates of amyloid accumulation* in the brain and higher rates of dementia. The relationship goes the other way as well: sleep disruption is an early symptom of dementia, creating a vicious cycle in which poor sleep accelerates the biological processes that further disrupt sleep.

The practical implications are significant. Getting seven to nine hours of sleep per night, treating sleep apnea when it's present, maintaining consistent sleep and wake times, and addressing sleep problems with medical help rather than accepting them as inevitable are all evidence-supported brain protection strategies.

The FINGER Trial: Multi-Domain Lifestyle Intervention

The Finnish Geriatric Intervention Study to Prevent Cognitive Impairment and Disability, known as the FINGER trial, was the first large randomized controlled trial to demonstrate that a *comprehensive lifestyle intervention could protect cognitive function in older adults* at elevated dementia risk.

The FINGER trial enrolled over 1,200 Finnish adults between ages 60 and 77 who were at elevated risk for dementia based on several factors. Participants in the intervention group received a comprehensive program addressing four domains simultaneously: nutritional guidance based on Finnish dietary guidelines, aerobic and strength exercise training, cognitive training, and management of cardiovascular and metabolic risk factors.

After two years, the intervention group showed significantly better cognitive performance than the control group, particularly in executive function, processing speed, and memory. The benefit was seen across multiple cognitive domains, suggesting that the comprehensive approach was more effective than targeting any single factor.

The FINGER model has since been replicated in multiple countries under the umbrella of the FINGERS World Wide Network, with adaptations for different cultural contexts. The results continue to support the principle that addressing multiple risk factors simultaneously produces cognitive benefit that single-factor interventions don't achieve.

The FINGER trial is the most important piece of evidence supporting a comprehensive lifestyle approach to dementia prevention. Its message is clear: exercise, eat well, train your brain, and manage your cardiovascular health, all at once. Not one at a time.

What You Can Start Today

Prevention research doesn't require you to wait for a prescription or a clinic appointment. The most evidence-supported prevention steps are available right now.

Blood pressure control: if you don't know your blood pressure, find out. If it's elevated, work with your physician to bring it to target. This is the single intervention with the strongest evidence for protecting brain health.

Physical activity: 150 minutes of moderate aerobic exercise per week. A 30-minute walk five days a week meets this target. If you're not doing this, start. The evidence is overwhelming.

Sleep: aim for seven to nine hours per night. If you snore heavily or wake exhausted, ask your doctor about sleep apnea testing. Treat it if it's present.

Hearing: if you suspect hearing loss, get a hearing test. If hearing aids are recommended, use them. The cognitive load of unmanaged hearing loss is a direct dementia risk factor.

Social engagement: maintain meaningful social connections. Join a group, volunteer, stay in touch with friends, attend community events. The brain that is socially engaged is a brain that continues to be challenged.

These are not revolutionary recommendations. But they are the ones backed by the best evidence. And evidence, not novelty, is what this book offers.

Diet and Brain Health

Diet is one of the modifiable risk factors with a growing body of evidence, and it deserves specific attention because the research has become considerably stronger in recent years.

The MIND diet, which stands for Mediterranean-DASH Intervention for Neurodegenerative Delay, was developed by nutritional epidemiologist Martha Clare Morris specifically to combine elements of the Mediterranean diet and the DASH diet that have the strongest evidence for brain health. The MIND diet emphasizes green leafy vegetables, other vegetables, berries, nuts, olive oil, whole grains, fish, beans, and poultry, while limiting red meat, butter and margarine, cheese, pastries and sweets, and fried or fast food.

In a large observational study published in 2015, people who adhered closely to the MIND diet had cognitive performance equivalent to someone 7.5 years younger, and had a significantly lower rate of Alzheimer's disease development, compared to those with the lowest adherence. While observational studies cannot prove causation, the size and consistency of these findings, combined with the plausible biological mechanisms involving inflammation, oxidative stress, and vascular health, make the MIND diet one of the most well-supported dietary recommendations for brain health.

Berries deserve specific mention. Blueberries and strawberries are among the richest dietary sources of flavonoids, compounds that have been shown in laboratory and epidemiological research to reduce oxidative stress and inflammation in the brain. Several observational studies have found that higher berry consumption is associated with slower cognitive decline in older adults.

Omega-3 fatty acids, found in fatty fish like salmon, sardines, and mackerel, are important structural components of brain cell membranes and have been associated with reduced inflammation. While omega-3 supplements have not shown consistent benefit in clinical trials for people who already have dementia, dietary omega-3 intake is associated with better cognitive aging in observational research.

The dietary message is not complicated: eat more plants, more fish, and more olive oil, and less ultra-processed food, red meat, and added sugar. This is consistent with what's good for heart health, metabolic health, and now, increasingly, brain health.

Cognitive Reserve: Building a Buffer

One of the most important concepts in dementia prevention is *cognitive reserve,* and it explains something that puzzles many people: why two people with apparently similar amounts of brain damage can show very different levels of cognitive function.

Cognitive reserve refers to *the brain's resilience and ability to compensate for damage.* A person with high cognitive reserve can tolerate more amyloid plaques, more tau tangles, and more neuronal loss before their cognitive function declines to the point of a dementia diagnosis. A person with lower cognitive reserve may show symptoms at a lower level of biological damage.

What builds cognitive reserve? Education is the factor with the strongest evidence. People who receive more years of formal education have higher cognitive reserve on average, which is why lower education in early life is one of the Lancet Commission's 12 modifiable risk factors. The mechanism is thought to involve the development of more complex and redundant neural networks that can compensate when damage occurs.

But education isn't the only factor. *Mentally stimulating work, lifelong learning, social engagement, bilingualism, and musical training* have all been associated with higher cognitive reserve. The principle is consistent: the more the brain is used and challenged across a lifetime, the more resilient it becomes.

This has important practical implications. It's never too late to invest in cognitive reserve. *Learning a new language, taking up a musical instrument, pursuing formal or informal education, staying in complex and stimulating social roles,* all of these activities may contribute to the brain's capacity to withstand and compensate for the changes that aging and disease bring.

Cognitive reserve doesn't prevent Alzheimer's disease. People with high cognitive reserve still get the disease.

Chapter 6 Summary

The Lancet Commission's 2020 analysis identified 12 modifiable risk factors that together account for approximately 40 percent of dementia cases worldwide. Key factors include high blood pressure in midlife, hearing loss, physical inactivity, social isolation, smoking, obesity, and depression. Cardiovascular health and brain health are deeply connected, and managing cardiovascular risk factors is brain protection. The discovery of the glymphatic system has revealed that sleep is essential for clearing amyloid from the brain, and chronic sleep deprivation is a significant risk factor. The FINGER trial demonstrated that comprehensive multi-domain lifestyle intervention protects cognitive function in older adults at elevated risk. The most evidence-supported prevention steps are available right now: blood pressure control, regular aerobic exercise, adequate sleep, hearing loss treatment, and social engagement.

Three Things You Can Do Right Now

1. Know your blood pressure. If you don't, get it measured this week. If it's above 130 over 80, talk to your doctor about what to do about it.

2. Schedule a hearing test if you have any hearing concerns at all. Hearing loss is the largest single modifiable dementia risk factor in midlife.

3. Get a realistic handle on your sleep this week. Track how many hours you actually sleep each night. If it's consistently under seven, that's a problem worth solving.

Chapter 7: Technology and Artificial Intelligence in Dementia Care

Technology is changing dementia care faster than almost any other part of the field.

Not just the technology covered in Book Two, the medication dispensers and GPS trackers and video calling tools that support daily life. Those are important, and they're covered there. What this chapter addresses is the frontier: *how artificial intelligence and advanced technology are changing how we diagnose dementia, how we run clinical trials, how we discover new drugs, and how we deliver care to people who otherwise wouldn't have access to it.*

Some of what's in this chapter is available right now. Some is a few years away. All of it is heading in the same direction: toward earlier detection, more personalized care, and a research process that moves faster than it ever has before.

AI-Assisted Diagnosis

Artificial intelligence, specifically machine learning algorithms trained on large datasets, is being applied to dementia diagnosis in several promising ways.

Retinal imaging is one of the most striking examples. The eye is an extension of the brain, and changes in the retinal blood vessels and nerve layers mirror what's happening in the brain's blood vessels and neurons. Several research groups have developed AI algorithms that can analyze retinal photographs, the same type of image taken during a standard eye examination, and identify patterns associated with Alzheimer's disease years before cognitive symptoms appear. If validated in large-scale studies, *this could make Alzheimer's screening as routine as a visit to the eye doctor.*

Speech and language analysis is another active area. The way people speak, their word choice, sentence structure, pauses, and fluency, changes subtly in the early stages of dementia. *AI systems trained on recordings of speech can identify these changes* with a sensitivity that exceeds human detection. Researchers have shown that speech analysis can identify people who will later develop dementia from recordings of their speech made years earlier.

Digital cognitive assessments, delivered on a tablet or smartphone, offer another avenue. These assessments can be administered more frequently and in more convenient settings than traditional neuropsychological testing, and AI analysis of the results can track subtle changes over time that might not trigger concern in a single assessment but represent meaningful trends across multiple measurements.

These AI-based tools are promising but not yet standard of care. Their validation in large, diverse populations is ongoing, and questions about accuracy, bias, and appropriate clinical use remain. But the direction is clear: earlier, more accessible, and more sensitive diagnosis is coming.

Wearable Sensors and Passive Monitoring

Changes in how people move, sleep, and go about their daily routines precede cognitive symptoms in dementia. Gait changes, altered sleep architecture, heart rate variability differences, and disruptions in daily activity patterns have all been associated with early dementia risk.

Wearable sensors, including smartwatches, fitness trackers, and specialized research devices, can capture these signals continuously and passively without requiring any deliberate effort from the person being monitored. AI algorithms can then analyze this continuous data stream to detect changes that might indicate early cognitive decline.

The digital biomarker approach, using wearable data as a biomarker for brain health, is being incorporated into clinical trials as both an eligibility criterion and an outcome measure. This allows trials to track changes in real-world behavior rather than relying solely on periodic in-clinic cognitive tests, potentially making trials more sensitive and reducing the number of participants needed.

For people living with dementia and their families, wearable monitoring offers safety and care benefits as well. Devices that detect falls, monitor sleep patterns, track physical activity, and alert caregivers to unusual patterns provide a layer of oversight that wasn't previously possible without constant human supervision.

AI in Drug Discovery

One of the most time-consuming and expensive parts of drug development is *identifying which molecules might be effective against a specific biological target.* Traditionally, this has involved screening millions of compounds through a slow, expensive, and largely manual process.

AI is radically accelerating this process. Machine learning algorithms can analyze the biological structures of known compounds, predict how new compounds might interact with target proteins, identify candidates that are likely to be both effective and tolerable, and prioritize which ones are most worth testing. In several cases, AI has identified drug candidates in months that would have taken years to find through conventional screening.

The implications for dementia research are significant. Multiple companies and research institutions are now using AI-driven drug discovery platforms specifically for neurodegeneration. The diversity of targets being explored, including amyloid, tau, neuroinflammation, synaptic function, and the gut-brain axis, means that the pipeline of potential new treatments is broader than it has ever been.

AI is also being used to analyze the vast datasets generated by clinical trials, patient health records, and biobank samples to identify patterns that humans would miss: subgroups of patients who respond better to certain treatments, early signals of treatment benefit or harm, and biological markers that predict who will progress faster or slower.

Telemedicine and Access to Care

One of the persistent inequities in dementia care is geographic: the specialists and research centers with the most expertise are concentrated in major metropolitan areas. Rural communities, smaller cities, and economically disadvantaged communities have far less access to the neurologists, neuropsychologists, and geriatric specialists who provide optimal dementia care.

Telemedicine is addressing this gap. The pandemic accelerated the adoption of telehealth across medicine, and for dementia care specifically, the evidence suggests that cognitive assessments, medication management consultations, and caregiver support can be delivered effectively by video visit for many patients.

Clinical trials are also increasingly incorporating telemedicine. Participants who can complete some visits remotely can enroll from locations that would otherwise be too far from a trial site. *This expands diversity in trial enrollment* and makes participation accessible to people who otherwise couldn't participate.

The integration of telemedicine with digital biomarkers from wearables and AI-assisted cognitive monitoring creates the possibility of continuous, accessible dementia care that doesn't require a person to travel to a specialist clinic every month. This is not fully realized yet, but it is the direction the field is moving.

Ethical Considerations

As powerful as these technologies are, they raise genuine ethical questions that deserve serious consideration.

Privacy is paramount. Continuous monitoring through wearables generates intimate data about a person's daily life, health, and behavior. Who owns that data? Who can access it? How is it protected? For people with dementia who may not fully understand what they've agreed to, these questions carry particular weight.

Algorithmic bias is a real risk. AI systems trained on data that underrepresents certain populations, including racial and ethnic minorities, may perform less accurately for those populations. Ensuring that the AI tools being developed for dementia work equally well across diverse groups is an ethical and scientific imperative.

The automation of care also raises questions. AI-powered tools and robotic assistants can extend care capacity, but *they are not substitutes for human connection.* The value of a human presence, a person who knows and is invested in the individual with dementia, cannot be replicated by a device. Technology should augment human care, not replace it.

These are conversations happening actively in the research and clinical communities, and they will continue as the technology evolves. Patients, families, and advocacy organizations have an important role in shaping how these technologies are developed and deployed.

Social Robots and Virtual Companions

One area of technology that is increasingly used in dementia care settings, and that raises particularly interesting questions, is social robotics: robots and virtual companions designed specifically to provide interaction, engagement, and emotional connection for people with dementia.

PARO, a robotic seal designed to respond to touch and sound with movements and sounds of its own, is the most well-known example. PARO has been studied in multiple clinical settings and has been shown to reduce agitation, improve mood, and increase social engagement in people with dementia. Some studies have found that PARO produces physiological calming effects, including reduced cortisol levels. It is used in *care facilities in Japan, Europe, and increasingly in the United States.*

AI-powered virtual assistants specifically designed for people with dementia, distinct from general-purpose assistants like Alexa, are also in development. These systems are designed with the specific cognitive and communication needs of dementia in mind: simplified interfaces, memory of previous conversations, reminders tailored to individual routines, and the ability to engage in reminiscence conversations using photographs and personal history.

The ethical dimension here is complex. For people with dementia who have diminished capacity to distinguish robotic companions from human ones, questions arise about authenticity and consent. Does a person with dementia in the later stages need to know that the 'friend' they're interacting with is a robot? Different ethicists and clinicians hold different views on this question, and there is no consensus.

What is clear is that social isolation and boredom are genuine harms for people with dementia, and that anything that safely reduces those harms improves well-being. Technology that serves that function has value, provided it is used as a complement to human connection and not as a substitute for it.

Chapter 7 Summary

AI is being applied to dementia diagnosis through retinal imaging analysis, speech pattern analysis, and digital cognitive assessments, all of which may enable earlier and more accessible detection. Wearable sensors capture continuous data on gait, sleep, and daily activity patterns that can detect early changes associated with dementia and provide safety monitoring for people living with the disease. AI is accelerating drug discovery by rapidly identifying and prioritizing drug candidates across a broader range of biological targets than was previously feasible. Telemedicine is improving access to dementia care and clinical trial participation for people in underserved geographic areas. These technologies raise important ethical questions about privacy, algorithmic bias, and the appropriate role of automation in care, questions that patients and families should be part of answering.

Three Things You Can Do Right Now

1. If your loved one with dementia uses a smartwatch or fitness tracker, ask their neurologist whether the data might be useful for tracking changes between visits. Some research centers are actively incorporating this data.

2. If distance from a specialist is a barrier, ask the primary care physician about telehealth referral options to dementia specialists. The availability of these services has expanded significantly since 2020.

3. Pay attention to news about AI in dementia research. The stories are moving fast. Alzheimer's News Today at alznewstoday.com provides accessible daily coverage of research developments.

Chapter 8: Stem Cells, Gene Therapy, and the Science of the Future

Some of the most exciting work in dementia research is happening at the frontier of what's currently possible.

The treatments in this chapter are not yet available in the clinic for most people. Some are in early clinical trials. Some are still in laboratory research. All of them represent directions the field is moving, and understanding them gives you a sense of what the next decade might bring.

A word of caution before we begin: the history of medicine is littered with treatments that looked spectacular in the laboratory and failed in human trials. This chapter presents the current state of promising research, not promises of cures. But the research is real, and the progress is genuine.

Stem Cell Research

Stem cells (also known as pluripotent cells) are cells that have the ability to *develop into many different types of specialized cells.* The idea of using stem cells to treat Alzheimer's disease has been around for decades, and it takes several different forms in current research.

One approach is cell replacement therapy: *using stem cells to generate new neurons that could replace the neurons lost in dementia.* This is appealing in concept but enormously complex in practice. The brain is not like a muscle or a skin layer that regenerates naturally. The neural circuits that encode memories, personalities, and knowledge built over a lifetime are extraordinarily specific and complex. *Simply adding new neurons to a damaged brain does not restore the circuits that were lost.*

A more promising near-term approach is using stem cells to deliver neuroprotective factors: substances that support the survival of existing neurons and promote the brain's own repair mechanisms. Clinical trials are testing stem-cell-derived therapies that inject cells engineered to release factors that reduce inflammation and support neuronal health.

Stem cells are also invaluable research tools. Scientists can now *take skin or blood cells* from a person with Alzheimer's disease and use them to generate neurons that carry the same genetic profile as that person's brain cells. These patient-derived neurons can be studied in the laboratory to understand what's going wrong at a cellular level and to test potential treatments on human cells without exposing patients to risk. This, as you can understand, is research-based work currently.

Gene Therapy

Gene therapy involves changing, adding, or silencing genes in a patient's cells to treat or prevent disease. For Alzheimer's disease, gene therapy approaches are being explored in several directions.

CRISPR-Cas9 gene editing is the most widely discussed gene-editing tool. CRISPR can be used to precisely edit the DNA sequence of a gene with unprecedented accuracy. In the context of Alzheimer's, researchers are exploring whether CRISPR can be used to edit the APOE4 gene to convert it into the less risky APOE3 form, or to modify genes involved in amyloid production or clearance. Early research in animal models and cell systems is underway, but clinical application in humans is years away.

Antisense oligonucleotides, which are molecules that can silence or reduce the activity of specific genes, are further along in development. Several antisense drugs targeting tau protein are in clinical trials for Alzheimer's and related conditions. These drugs don't edit the DNA itself but instead intercept the molecular message between the gene and the protein it produces, reducing how much tau protein is made.

Gene therapy approaches using viral vectors to deliver protective genes into the brain are also in early clinical trials. Some trials are testing the delivery of genes that produce neuroprotective factors directly into the brain tissue, aiming to provide sustained protection that outlasts a single drug dose.

Targeting Neuroinflammation

The growing recognition that neuroinflammation is a major driver of Alzheimer's disease has opened up an entirely new category of drug development targets.

Several drugs designed to modify the activity of microglia, the brain's immune cells, are in clinical trials. Some aim to reduce the harmful inflammatory response that damaged microglia produce. Others aim to restore microglial function so the cells can more effectively clear amyloid and other waste products from the brain.

The TREM2 protein, which regulates microglial function, is one of the most actively studied targets in this space. Mutations in the TREM2 gene are associated with significantly increased Alzheimer's risk, and drugs designed to activate or supplement TREM2 function are in early clinical development.

Complement system inhibitors are another area of interest. The complement system is part of the immune system that can be activated in the brain in ways that contribute to neuronal damage. Drugs that modulate complement activation in the brain may protect neurons from immune-mediated damage.

This neuroinflammation-targeted drug development represents a fundamentally different approach from the amyloid-focused therapies: instead of clearing the garbage that's causing the fire, it's trying to fix the fire department.

The Microbiome-Brain Connection

One of the most surprising areas of emerging research is *the connection between the gut microbiome and brain health.* The gut microbiome is the community of trillions of microorganisms that live in the digestive system. Research in the last decade has revealed that the gut and the brain communicate through multiple pathways, including the vagus nerve, immune signaling, and the production of metabolites that affect brain function.

Studies in both animal models and human populations have found associations between the composition of the gut microbiome and the risk of Alzheimer's disease. People with Alzheimer's disease tend to have *different microbiome compositions than cognitively healthy people of similar ages.* Whether these differences are a cause, a consequence, or both is still being determined.

Research is exploring whether modifying the gut microbiome through diet, probiotics, fecal microbiota transplantation, or other means can affect brain health. Several small clinical trials are underway. This is *very early-stage science*, but the biological mechanisms are plausible and the research is accelerating.

The most practical implication of microbiome research for people right now is *dietary*: a diet rich in fiber, fermented foods, and plant diversity supports a healthy and diverse gut microbiome. This is consistent with the dietary patterns already recommended for brain health, like the MIND diet, which suggests these patterns may benefit the brain through multiple pathways simultaneously.

Precision Medicine: Treating Subtypes Differently

One of the most significant shifts in Alzheimer's research in recent years is the recognition that *Alzheimer's disease is not one uniform biological entity.* Different people with Alzheimer's have different combinations of biological changes, different rates of amyloid and tau accumulation, different degrees of neuroinflammation, and different patterns of genetic and metabolic risk.

Precision medicine, sometimes called personalized medicine, applies the principle that different biological subtypes of Alzheimer's may respond better to different treatments. The same drug that produces significant benefit in a patient whose disease is driven primarily by amyloid accumulation may have little effect in a patient whose disease involves predominantly tau pathology or neuroinflammation.

This thinking is already influencing clinical trial design. Trials are increasingly stratifying participants by biomarker profile, treating people who meet specific biological criteria *rather than just clinical diagnostic criteria*. Donanemab's trial found greater benefit in participants with lower tau burden, a finding that suggests subtype matters for treatment outcome.

Precision medicine in Alzheimer's disease is still in its early stages, but it represents the direction the field is moving: away from treating everyone with Alzheimer's the same way, and toward treatment approaches tailored to the specific biological profile of each individual.

Senolytic Drugs and the Biology of Aging

One of the most intriguing new directions in Alzheimer's research connects the disease to the fundamental biology of aging, specifically to a phenomenon called cellular senescence.

Senescent cells are cells that have stopped dividing but haven't died. Instead, they persist in tissues and organs, producing a cocktail of inflammatory molecules that damage surrounding cells. As people age, senescent cells accumulate throughout the body, including in the brain, where they may contribute to neuroinflammation and neurodegeneration.

Senolytic drugs are compounds that selectively destroy senescent cells, potentially reducing the chronic low-grade inflammation that aging cells produce. In animal studies, senolytic drugs have shown promising effects on brain health: reducing amyloid and tau pathology, decreasing neuroinflammation, and in some models, improving cognitive performance.

Several senolytic drugs, including combinations of quercetin and dasatinib, are in early clinical trials for Alzheimer's disease and related conditions. These trials are in early stages, and it's too soon to know whether the effects seen in animal studies will translate to humans. But the approach is scientifically plausible and represents a genuinely new angle on dementia biology.

The broader implication of senolytic research is that Alzheimer's disease may be understood, in part, as an accelerated aging process in the brain, and that interventions targeting the fundamental biology of aging might provide new treatment avenues that complement the amyloid- and tau-focused approaches currently dominant in the field. This is an area to watch as the science develops.

Chapter 8 Summary

Stem cell research for Alzheimer's is exploring both cell replacement therapy and the use of stem cells to deliver neuroprotective factors, with patient-derived neurons serving as invaluable research tools. Gene therapy approaches include CRISPR-based editing of risk genes, antisense oligonucleotides targeting tau, and viral vector delivery of neuroprotective genes, most of which are in early trial stages. Neuroinflammation-targeted drug development is pursuing drugs that modify microglial activity, TREM2 activation, and complement system modulation. The gut microbiome is connected to brain health through multiple biological pathways, and research is exploring whether microbiome modification can affect Alzheimer's risk. Precision medicine is shifting the field toward treating different biological subtypes of Alzheimer's with different approaches, already influencing clinical trial design.

Three Things You Can Do Right Now

1. Support your gut microbiome through diet by increasing fiber, fermented foods like yogurt and kefir, and variety in the plant foods you eat. This is aligned with both gut health and brain health research.

2. When reading about gene therapy or stem cell research for Alzheimer's, check whether the research was done in human clinical trials, in animal models, or in cell systems only. The distinction matters enormously for how near-term the implications are.

3. The Alzheimer's Drug Discovery Foundation at alzdiscovery.org publishes regular updates on the drug development pipeline in accessible language. Add it to your reading lists.

Chapter 9: The Role of Public Policy and Advocacy

The pace of dementia research is not determined only by scientists.

It's determined by funding, and funding is determined by policy, and policy is shaped by advocacy. The billions of dollars now flowing into dementia research didn't arrive because scientists asked nicely. *They arrived because people living with dementia, their families, their caregivers, and the organizations that represent them made the case, year after year, to the people who control the public purse.*

This chapter explains the policy landscape for dementia care and research, and it invites you to be part of shaping it. Not as a passive beneficiary of what policy produces. As an active participant in making it happen.

The National Plan to Address Alzheimer's Disease

In 2011, Congress passed the National Alzheimer's Project Act, which required the federal government to create and maintain a comprehensive national plan for Alzheimer's disease. The National Plan to Address Alzheimer's Disease has been *updated annually since then,* setting specific goals for research, care, caregiver support, and public awareness.

The most significant goal of the National Plan is the commitment to effectively prevent and treat Alzheimer's disease. While that goal has not yet been fully achieved, the plan has focused federal attention and resources on dementia in ways that have had real consequences. NIH funding for Alzheimer's research increased from approximately $600 million in 2015 to over $3.5 billion by 2024, driven in large part by advocacy-informed congressional action.

The National Plan also *sets goals for improving dementia care quality, supporting caregivers, and increasing public awareness.* Progress on these goals is tracked and published annually, creating accountability for federal action.

Staying informed about the National Plan and its implementation is one way to track whether commitments are being kept and to identify where advocacy is needed most.

The FDA and Dementia Drug Approval

The FDA's role in dementia treatment is both a gatekeeper and a shaper of what research gets done. The FDA's approval standards, and how it applies them to Alzheimer's drugs, have been the subject of significant controversy and important evolution.

The approval of lecanemab, the first disease-modifying Alzheimer's drug, went through the FDA's traditional approval process, which requires demonstration of clinical benefit, meaning actual improvement or slowing of decline in patients, not just a biological marker change. This was a higher bar than the accelerated approval that had been controversially applied to aducanumab.

The FDA's Breakthrough Therapy designation, which expedites the development and review of drugs for serious conditions, has been applied to several Alzheimer's drug candidates. This designation allows for more intensive FDA guidance during development and speeds the review process without lowering the evidence bar.

Patient advocacy organizations, including the Alzheimer's Association, provide formal input into FDA processes through public comment periods, advisory committee hearings, and direct advocacy. The patient perspective on what constitutes meaningful benefit and acceptable risk is explicitly part of the FDA's deliberation. This is one of the ways that people affected by dementia have real influence over the regulatory process.

Medicare Coverage and the Battle for Access

For many patients, the most consequential policy question about a new Alzheimer's treatment isn't whether it's approved. *It's whether Medicare will pay for it.*

The coverage dispute over aducanumab was one of the most contentious in Medicare's history. The Centers for Medicare and Medicaid Services initially restricted coverage to people enrolled in clinical trials, effectively making the drug inaccessible to most Medicare beneficiaries despite FDA approval. The drug was ultimately withdrawn from the market, but the coverage battle set a precedent that shaped how Medicare approached the next generation of Alzheimer's drugs.

For lecanemab and donanemab, Medicare coverage improved significantly, with both drugs covered for eligible Medicare beneficiaries who meet specific criteria. However, *cost-sharing requirements, prior authorization requirements, and the need for specialized monitoring* all create real barriers to access that advocacy organizations continue to work to address.

The Medicare coverage of dementia drugs is a live and evolving policy issue. The Alzheimer's Association maintains current information about coverage at alz.org and advocates actively for improved access. Contacting your congressional representatives about Medicare coverage for Alzheimer's treatments is one of the most direct ways to contribute to this advocacy.

How to Advocate

Advocacy doesn't require traveling to Washington or attending formal hearings. There are accessible, meaningful ways to participate from anywhere.

Contact your elected representatives. The Alzheimer's Association's advocacy page at *alz.org/advocacy* provides tools to contact your senators and congressional representatives about specific dementia-related legislation. It takes ten minutes. It matters because representative offices track constituent contacts and they influence what legislators prioritize.

Participate in Alzheimer's Association advocacy days. The national Advocacy Forum in Washington each year, as well as state-level advocacy days, brings people affected by dementia directly to legislators' offices. These events are open to everyone and provide training and support for first-time advocates.

Share your story. Personal stories are among the most powerful tools in advocacy. Legislators, researchers, and the public are moved by the human reality of dementia in ways they are not moved by statistics. If you're willing to share your experience, the Alzheimer's Association and other organizations can help you use it effectively. in the past, one of the most effective advocates for Alzheimer's research was the daughter of actress Rita Hayworth, who had developed Alzheimer's disease. The wife of one of New York state's largest automobile sellers contributed millions to the cause and appeared regularly in the media.

Support research funding organizations. Financial contributions to the Alzheimer's Association, the Alzheimer's Drug Discovery Foundation, the Alzheimer's Research Foundation, and other research-funding organizations directly support the research pipeline. These organizations also do the advocacy work that expands public funding.

The International Dimension

Dementia is a global crisis. The World Health Organization estimates that approximately 55 million people worldwide live with dementia, and that number is projected to reach 139 million by 2050. The burden falls disproportionately on low and middle-income countries, which will account for the majority of future dementia cases but have far fewer resources for research and care.

International research collaboration is increasingly important for the speed and scale of progress. The FINGERS World Wide Network, mentioned in the previous chapter, is one example of how clinical trials can be coordinated across countries to produce faster and more generalizable results. Large biobanks that share data across institutions and borders are allowing researchers to analyze patterns across populations that no single country could assemble alone.

The global nature of dementia also means that research done anywhere in the world may benefit you. A discovery in a laboratory in Japan, a clinical trial result from the United Kingdom, and a prevention study from Finland: all of these contribute to the global understanding that shapes treatment everywhere.

Knowing this is useful because it helps explain why *international news about dementia research is worth following.* The Alzheimer's Association International Conference, held annually, is *the largest gathering of dementia researchers in the world* and produces news coverage that reflects the global state of the field.

The Economic Argument for Investment

Beyond the human case for dementia research investment, there is a compelling economic argument that has become increasingly prominent in advocacy efforts.

The Alzheimer's Association estimates that the total cost of Alzheimer's and other dementias in the United States, including direct medical costs, long-term care costs, and the value of unpaid caregiver time, *exceeds one trillion dollars annually and is projected to grow dramatically as the population ages.* This cost is borne by patients, families, insurance systems, and government programs, particularly Medicare and Medicaid.

Every year that effective treatment is delayed is a year of enormous human and economic cost. Conversely, a treatment that delays the onset of Alzheimer's disease by even five years would reduce the prevalence of the disease by approximately 57 percent over the following decades, because people who die before the disease would have become symptomatic would never become cases.

The return on investment for Alzheimer's research is therefore staggering, even at relatively modest effectiveness levels. A dollar invested in Alzheimer's research today, if it produces even a partial treatment, could save many multiples of that dollar in long-term care costs over the coming decades.

This economic argument has been used effectively by the Alzheimer's Association and other advocacy organizations to make the case for sustained federal research investment. It resonates with legislators who might otherwise focus on the immediate costs of research rather than the much larger costs of the disease itself. Understanding this argument makes you a more effective advocate when you contact your representatives.

Health Disparities and Who Gets Left Behind

Any honest account of dementia research and policy must acknowledge that the benefits of research advances have not been and are not currently reaching all populations equally. *Health disparities in dementia are significant,* and addressing them is both a moral imperative and a scientific one.

African Americans are approximately twice as likely to develop Alzheimer's disease as white Americans, and Hispanic Americans are approximately one and a half times more likely. These elevated rates are not primarily driven by genetics. *They are driven by higher rates of the modifiable risk factors discussed in Chapter Six:* cardiovascular disease, hypertension, diabetes, and the social determinants of health that contribute to them, including socioeconomic inequality, chronic stress, and reduced access to preventive care.

Participation in clinical trials has historically underrepresented these communities. The trials that led to lecanemab and donanemab approval enrolled populations that were predominantly white and had higher-than-average levels of education. This means we have less certainty about how these drugs perform in the broader, more diverse population that will eventually receive them.

When I was working as the National Clinical Monitor for the trial of Cognex in the United States at Mount Sinai medical center in New York City, we was fortunate to have a physician from China join our team. This man went to New York City's Chinatown every day, interviewed and evaluated people in that population, and included them in our clinical trials. Without him, we would not have been able to use that knowledge in our research. He enrolled literally hundreds of patients.

The scientific community has recognized this as a problem and has made increasing diversity in clinical trial enrollment an explicit priority. The Alzheimer's Clinical Trials Consortium and several NIH-funded programs are specifically designed to recruit and support the participation of underrepresented communities in dementia research.

From an advocacy standpoint, the disparities in both disease burden and research participation are issues worth raising with legislators and policymakers. Ensuring that federally funded research reaches and benefits all populations, not just those already advantaged by socioeconomic circumstances, is a matter of both justice and scientific rigor.

If you or someone you love is from a community that has been underrepresented in Alzheimer's research, consider that participating in clinical trials is one of the most direct ways to change this. Your participation contributes data that makes the research more applicable to everyone who looks like you.

Chapter 9 Summary

The pace of dementia research and the accessibility of new treatments are shaped by policy decisions that are in turn shaped by advocacy. The National Plan to Address Alzheimer's Disease has driven significant increases in federal research funding since 2011. The FDA's approval standards for dementia drugs have evolved, with traditional approval based on clinical benefit now established for the newest disease-modifying treatments. Medicare coverage for Alzheimer's treatments is an active and evolving policy issue. Effective advocacy includes contacting elected representatives, participating in Alzheimer's Association advocacy events, sharing personal stories, and supporting research funding organizations. Dementia is a global crisis that requires international collaboration, and research advances anywhere in the world benefit patients everywhere.

Three Things You Can Do Right Now

1. Go to alz.org/advocacy and use the tool to contact your

senators and congressional representative about Alzheimer's research funding and treatment access. It takes ten minutes.

2. Consider registering for your state's Alzheimer's Association advocacy day. First-time advocates are welcome and supported. Find your local chapter at alz.org.

3. Follow Alzheimer's Association advocacy news to stay current on the policy issues most affecting dementia care and research access.

Chapter 10: Holding Hope

I want to end this series the way it has to end: with both honesty and hope.

Not false hope. Not the kind of hope that dismisses the reality of what dementia is and what it costs. The people who have read all four of these books have a clear-eyed understanding of that cost. They don't need it minimized.

But real hope. The kind that is grounded in evidence, that looks at what has actually changed in the last decade, and that makes the reasonable claim: we are not where we were, and where we're going is better than where we've been.

What Hope Is Not

Hope in serious illness is often misunderstood in ways that make it less useful than it could be.

Hope isn't the belief that everything will work out the way you want it to. It isn't the expectation of cure, or the guarantee that the new treatment will work for this particular person, or the certainty that the disease will not progress. People who define hope that way are setting themselves up for a fall when reality doesn't match the expectation.

Hope is not denial. It is not pretending that the hard parts aren't hard, or that the losses aren't losses. People who conflate hope with positive thinking often find themselves unable to grieve what needs to be grieved, or unable to make the practical plans that the reality of the disease requires.

And hope is not passive. It is not waiting for someone else to fix the problem, or expecting that the future will be better without any effort to make it so.

What Hope Is

Real hope, in the context of a serious and progressive disease, is the informed belief that the future is genuinely uncertain and that meaningful positive outcomes remain possible.

It includes the realistic assessment that things have changed. The person diagnosed with Alzheimer's disease today faces a different medical landscape than the person diagnosed ten years ago. They have access to treatments that slow the disease. They have blood tests that can confirm the diagnosis earlier. They have a research pipeline that is more active and better funded than at any previous point in history. The landscape is genuinely different.

Real hope includes the recognition that what you do matters. The lifestyle choices described in Chapter Six, the engagement with medical care described throughout this series, the decision to participate in a clinical trial if eligible: these are not small things. They are choices that affect outcomes, and making them is an expression of active hope.

Real hope includes the capacity to hold both difficulty and possibility at the same time. To know that today is hard, and that tomorrow might be better. To grieve what has been lost, and to remain invested in what is still possible. This is not a contradiction. It is the most honest way to live with a progressive disease.

The Ten-Year Horizon

Researchers and clinicians in the Alzheimer's field are, by scientific necessity, cautious about predictions. The history of the field includes too many moments of premature optimism to make specific promises about timelines credible.

But there is a reasonable case for what the next ten years might hold, based on what is currently in the pipeline and what the pace of progress suggests.

In the next several years, it is likely that additional disease-modifying treatments will receive FDA approval. Anti-tau drugs in current Phase 3 trials are expected to report results. Some may show benefit, adding tau-targeted therapy to the treatment arsenal. Neuroinflammation-targeted drugs are not far behind in the pipeline.

Earlier intervention will become possible as blood biomarkers are refined and standardized. The ability to detect Alzheimer's pathology from a simple blood test in a primary care setting will change who gets evaluated, when they get evaluated, and how early treatment can begin. The window for effective disease modification will open wider as earlier-stage patients can be reliably identified.

Prevention trials currently underway will report results in the next several years. If multi-domain lifestyle intervention trials confirm and extend the FINGER results, evidence-based prevention programs may become a standard part of primary care for middle-aged and older adults at elevated risk.

Combination therapies, attacking the disease from multiple biological angles simultaneously, are being designed and will enter large-scale trials. If they work, they may produce substantially greater slowing than any single agent alone.

None of this is certain. The history of the field demands humility about predictions. But the direction of the evidence, and the scale of the investment, and the proof of concept that disease modification is possible all point toward a decade that will look meaningfully different from the decade we just left.

The People Making It Happen

Behind every clinical trial result, every approved drug, every blood test that can now detect Alzheimer's changes before symptoms appear, there are people.

There are researchers who worked in dementia for decades during the long period of failures, who didn't leave the field when the drugs kept failing, who kept asking the same question from different angles until the answers started coming. People like Bart De Strooper, whose basic science work on the proteins that produce amyloid was foundational to the drugs that now exist. People like Reisa Sperling, whose prevention trials are working to establish what early intervention can achieve. People like the team at the Banner Alzheimer's Institute, who are studying families with deterministic Alzheimer's genes to understand the disease before symptoms begin.

And there are the people who enrolled in the trials. The thousands of men and women who agreed to receive an infusion of an experimental drug every two weeks for 18 months, who submitted to repeated brain scans and blood tests, who kept diaries and came back to the clinic appointment after appointment, knowing they might be in the placebo group, knowing the drug might not work, doing it anyway because they understood that someone has to go first.

Those people are the reason that lecanemab and donanemab exist. They are the reason the next generation of drugs is moving through trials right now. The ones who come after them will benefit from what they did.

What You Can Do Right Now

This is not a series that ends with advice to wait for someone else to solve the problem. Here is what you can do right now, wherever you are.

1. If you or someone you love has been recently diagnosed: get a comprehensive evaluation at a center with dementia specialty expertise. Ask specifically about clinical trial eligibility. Engage with the full range of lifestyle strategies in this series, not just medication. Connect with the Alzheimer's Association for support and information.

2. If you have a family history of Alzheimer's: manage your modifiable risk factors aggressively, starting today. Know your blood pressure and what to do about it. Exercise. Sleep. Treat hearing loss. Stay socially connected. Look into prevention trials at trialmatch.alz.org.

3. If you are a caregiver: take care of yourself with the same seriousness you bring to caring for someone else. Get respite. Get

support. Know your own health. The Alzheimer's Association helpline at 1-800-272-3900 is there for you around the clock. Too often, caregivers tend not to attend to their own needs and feel guilty if they take some time off. Remember this: you can't care for someone else if you aren't caring for yourself.

4. If you want to support the work: donate to the Alzheimer's Association or the Alzheimer's Drug Discovery Foundation. Advocate for research funding. Share your story. Vote for people who fund science.

5. And if you have made it through all four books: you know more than most people about one of the most important medical challenges of our time. That knowledge is valuable. Use it.

A Final Word

I began this series with a simple commitment: to tell the truth. About what dementia is, about what it costs, about how to live with it, about how to care for someone who has it, and about where research is going and why there is reason to believe the future will be different from the past.

The truth includes hard things. Dementia is still progressive and still, for most people, incurable. The treatments that now exist slow it down; they don't stop it. The lifestyle interventions that reduce risk don't eliminate it. The caregiving that love produces is still exhausting. The grief that accompanies all of this is still real.

But the truth also includes this: **we have never been closer.** The generation of researchers working today is building on decades of science in ways that are producing real results. The people living with dementia today, and caring for people who are, are living through a transition. They arrived at this disease in one era of medicine and they are watching the field move toward something better.

That matters. It doesn't fix everything. But it matters.

Thank you for reading. Take care of yourself. Take care of the people you love. And keep going.

A Note on the Series as a Whole

This is the fourth book in the *Living With Dementia* series, and if you've read all four, you've covered an enormous amount of ground.

Book One, When Memory Fades, was about the beginning: understanding the diagnosis, the different types of dementia, how doctors evaluate it, what the stages look like, and how to start making sense of everything from a legal, financial, and emotional standpoint.

Book Two, Day by Day, was about the full toolkit: medications old and new, cognitive therapy, music, creative therapies, exercise, speech and language, mental health, behavioral strategies, technology, and how to build a personal care plan from all of it.

Book Three, Standing Beside Them, was written specifically for caregivers: the people providing care day after day, often invisibly, often without adequate support. It covers communication, daily care tasks, managing behavioral symptoms, burnout, respite, legal and financial navigation, family dynamics, grief, and what comes after caregiving ends.

And this book, *Tomorrow's Promise*, has been about where research is going and what there is to genuinely hope for.

Together, the series forms a complete map of the dementia landscape from diagnosis through research frontier. No single book in the series tells the whole story. But each one tells the part of the story that matters most to the person who needs it.

If there is someone in your life who could use one of these books, share it. The information in these pages exists to help people, and it can only help the people who find it.

Chapter 10 Summary

Real hope in serious illness is not the expectation of cure or the denial of difficulty. It is the informed belief that meaningful positive outcomes remain possible, grounded in what has actually changed. The medical landscape for Alzheimer's disease today is genuinely different from what it was ten years ago: disease-slowing treatments exist, earlier and less invasive diagnosis is possible, and the research pipeline is more active and better funded than ever before. The people who have made these advances possible include decades of dedicated researchers and thousands of clinical trial participants who went first. Anyone touched by dementia can contribute: by engaging fully with medical care and lifestyle strategies, by participating in clinical trials if eligible, by managing modifiable risk factors, by supporting research funding and advocacy, and by staying current with the rapidly evolving science.

Three Things You Can Do Right Now

1. Bookmark the Alzheimer's Association website at alz.org, the National Institute on Aging at nia.nih.gov, and Alzheimer's News Today at alznewstoday.com. Check them regularly. The story is moving fast.

2. Choose one action from this series, whether a lifestyle change, a medical appointment, a support group contact, or an advocacy step, and commit to doing it in the next seven days.

3. Share this book with one person who doesn't know how much the landscape of dementia has changed. The knowledge that hope is grounded in real evidence is itself a gift.

Chapter 11: Access, Cost, and Getting the Treatments That Exist

Here is a situation that plays out every week in doctors' offices across the country.

A person with early Alzheimer's disease hears about lecanemab or donanemab from a family member who read about them online. They bring the article to their appointment and ask their doctor about it. And their doctor has to explain that even though these drugs are FDA-approved and have been shown to slow the disease, actually receiving them *is not as simple as picking up a prescription.*

The drugs require infusion at a certified center, not a pharmacy. They require a confirmed amyloid PET scan or blood biomarker test before they can be prescribed. They require regular monitoring brain scans for the first year of treatment. *They cost tens of thousands of dollars per year.* Medicare coverage took years to fully work out. *Many neurology practices still don't offer them.* And not every patient is a good candidate even if they want them.

None of that means the treatments aren't worth pursuing. They are, for the right people. But it does mean that getting access to them requires knowledge, persistence, and often significant advocacy. This chapter gives you that knowledge.

Understanding the Gap Between Approval and Access

When the FDA approves a drug, it means the drug has been shown to be safe enough and effective enough to be available to patients. It does not mean the drug is immediately available to every patient who needs it, through every doctor who sees that patient, at a cost everyone can afford.

The gap between approval and access is real, and it's especially wide for the new anti-amyloid drugs. Several things have to align for a patient to actually receive treatment.

First, the treating physician has to be at a practice that can prescribe and monitor these drugs. That typically means a neurology practice, a memory clinic, or an academic medical center with the specialized staff and equipment required. *Primary care physicians generally cannot initiate these treatments on their own.*

Second, the patient has to have a confirmed diagnosis of early-stage Alzheimer's disease with documented amyloid pathology. That requires a blood biomarker test or amyloid PET scan, whichever the treating physician chooses. *Not every community has access to PET scanning, and the blood tests, while increasingly available, need to be ordered and interpreted correctly.*

Third, the infusion itself has to be administered at a site with the trained staff and monitoring protocols required. Infusion centers have expanded in recent years, but rural and underserved areas remain significantly behind.

Knowing that these barriers exist is the first step toward working through them.

What These Treatments Actually Cost

Lecanemab, sold as Leqembi, was *initially priced at approximately $26,500 per year* for a patient of average weight. Donanemab, sold as Kisunla, came in at a similar price point. Both are administered as intravenous infusions every two weeks, which adds facility and nursing costs on top of the drug price.

On top of the drug cost, patients need regular MRI brain scans, particularly during the first year of treatment when the risk of amyloid-related imaging abnormalities is highest. *Those scans can cost several hundred to several thousand dollars each* without insurance, depending on the facility and region.

For a person with no insurance or coverage gaps, these costs are *out of reach for most households.* This is why understanding your insurance coverage, and knowing where to look for help, is essential.

Medicare, Medicaid, and Private Insurance Coverage

Medicare, the federal health insurance program for people 65 and older and those with certain disabilities, has worked through its coverage decisions for the new anti-amyloid drugs. After initially limiting coverage to patients enrolled in clinical trials or registries, the Centers for Medicare and Medicaid Services *expanded coverage for lecanemab and donanemab in 2024 for patients who meet clinical criteria.*

Under Medicare coverage, patients are still responsible for their share of costs including deductibles and copayments, which can be significant. Medicare supplemental insurance, often called Medigap, can reduce those out-of-pocket costs substantially. If you or your loved one has both Medicare and a Medigap policy, understanding exactly what that policy covers before beginning treatment is important.

Medicaid coverage for these drugs varies by state. Some states have moved more quickly than others to cover them for eligible low-income patients.

Private insurance coverage also varies. Many private insurers cover these drugs for patients who meet clinical criteria, though prior authorization requirements can create delays. Prior authorization means the insurance company wants to review whether the treatment is medically necessary before it will pay. Your doctor's office can submit the documentation, but it's useful to know this process exists and to follow up on it actively.

The key action: before your first infusion, *call your insurance company directly and ask them specifically what your out-of-pocket costs will be for the drug itself,* the infusion facility fees, and the required monitoring MRI scans. Get that information in writing if you can.

Patient Assistance Programs

Both Eisai (the maker of lecanemab) and Eli Lilly (the maker of donanemab) have *patient assistance programs* designed to help people who cannot afford the drug costs even with insurance. These programs are available to patients who meet income and insurance eligibility criteria, and they can significantly reduce or eliminate drug costs for qualifying patients.

To find out if you or your loved one qualifies, *contact the manufacturer directly*. Lecanemab's patient support program is called *LEQEMBI CARE*, and Eli Lilly has a program called *Lilly Cares*. Your neurologist's office may also have a social worker or patient navigator who can help you access these resources.

The Alzheimer's Association also maintains a list of financial assistance resources for people with dementia and their families. Their helpline at 1-800-272-3900 is staffed 24 hours a day and can help you figure out what financial support may be available in your situation.

Finding a Treatment Center

Because these drugs require infusion at a specialized center, finding where to receive treatment is a practical challenge for many families, particularly those who don't live near a major medical center.

The best starting point is to ask your neurologist or memory care specialist whether their practice offers these treatments, and if not, whether they can refer you to a practice that does. The Alzheimer's Association's website at alz.org has a resource called the *Care Finder*, which can help identify dementia specialists and memory clinics in your area.

Academic medical centers, university hospitals, and large health systems with neurology departments are the most likely to have established infusion programs for these drugs. If you live far from such a center, it is worth asking whether some of the monitoring appointments can be done locally, with the infusion itself done less frequently at a more distant center.

The distance and logistics involved in receiving these treatments are real obstacles, and it's worth having an honest conversation with your doctor about whether the access burden makes sense for your specific situation, health status, and stage of disease.

If You Are Not a Candidate Right Now

Not everyone who has a dementia diagnosis is a good candidate for the current anti-amyloid treatments. *These drugs are approved for people with mild cognitive impairment or early-stage Alzheimer's disease* with confirmed amyloid pathology. People with more advanced disease, people who take blood thinners, people who carry two copies of the APOE4 gene and have a higher risk of a specific type of side effect, and people with certain other medical conditions may not be candidates.

If your doctor has told you that these treatments are not appropriate for you right now, *that does not mean your options are exhausted.* The lifestyle interventions described in Chapter 6 remain powerful, evidence-based tools. The current treatment medications for managing symptoms remain available and useful. Clinical trials for different drug targets, some of which may be more appropriate for your stage of disease, are enrolling.

The field is also moving fast. Drugs that target tau rather than amyloid are in late-stage trials and may offer a different option for people who are not good amyloid-treatment candidates. Staying engaged with your care team about what is available and what you might qualify for is the best approach.

Chapter Summary

The new anti-amyloid drugs for Alzheimer's disease represent a genuine advance, but getting access to them requires navigating a set of practical barriers: specialist care, biomarker confirmation, infusion centers, insurance approval, and cost. Medicare and most private insurers now cover these drugs for eligible patients, and patient assistance programs exist to help with costs that insurance doesn't cover. The Alzheimer's Association helpline at 1-800-272-3900 is a reliable starting point for questions about both clinical access and financial assistance. People who are not currently candidates for these treatments still have meaningful options in lifestyle, symptom management, and clinical trial participation.

Three Things You Can Do Right Now

1. Call your neurologist or primary care doctor and ask directly: Am I eligible for lecanemab or donanemab, and if not, why not and when might I be? Write down the answer so you can track it over time.

2. Contact your insurance company and ask specifically about coverage for anti-amyloid infusion therapy for Alzheimer's, including what your out-of-pocket maximum would be. Do this before you need the information, not after.

3. Visit alz.org or call 1-800-272-3900 and ask about the Care Finder and about financial assistance programs for Alzheimer's treatments. A real person will answer any time of day or night.

Chapter 12: Staying Informed: How to Follow the Science as It Moves

There is no shortage of information about dementia and Alzheimer's research. *There is a shortage of reliable information,* and an abundance of the other kind.

On any given week, you can find a headline claiming scientists have discovered a breakthrough treatment, followed the next day by a different headline saying the same approach failed. You can read breathless coverage of a mouse study suggesting a cure is imminent, and you can read, buried in paragraph twelve of a different story, that the same approach has failed in human trials five separate times. You can find supplements, diets, and devices marketed with language that sounds scientific but isn't.

For someone who is living with dementia, or caring for someone who is, navigating this landscape is not just exhausting. *It can lead to real harm: money spent on things that don't work, false hope followed by crushing disappointment, or real treatments not pursued because someone read something discouraging online.*

This chapter is a practical guide to staying genuinely informed. Not just informed about what the headlines say. Informed about what the science actually shows.

Why Medical News Is Harder to Read Than It Looks

Science is a slow process that moves in small steps. Medical journalism, by contrast, tends to report on individual studies *as if each one settles a question permanently.* The result is *a pattern that frustrates scientists and confuses readers*: a study comes out suggesting coffee protects against Alzheimer's, then another one says it doesn't, then a third says it depends on how much you drink and what your genetics are.

Each of those studies may be entirely valid. The problem is that no single study proves anything by itself. What matters is the accumulation of evidence across many studies, the quality of that evidence, and whether findings have been replicated by independent research teams.

A few questions can help you read any health story more critically.

1. Was this a study in animals or people? Animal studies are an important step in research, but they frequently don't translate to human outcomes.

2. How many people were in the study? Small studies often produce results that don't hold up in larger populations.

3. Was this a randomized controlled trial, which is the strongest study design, or an observational study, which can show

associations but not prove causation?

4. Who funded the research, and is there a conflict of interest to be aware of?

There is one thing we have now that we never had before: *artificial intelligence programs can summarize* whatever you are trying to read, whether it is a disclosure statement, an article, or anything else. Check to see which ones you might have on your computer. Many of the internet browsers will also have an AI summary tool on them.

You don't need to answer all of those questions for every story you read. You just need to know they exist, so that you can treat individual headlines with appropriate skepticism and reserve your conclusions for patterns of evidence across many studies.

Recognizing Misinformation About Dementia

Certain kinds of content about dementia are reliably unreliable. Recognizing them can save you time, money, and grief.

Be skeptical of any product that claims to prevent, treat, or cure Alzheimer's or any other dementia. As of the time this book was written, no supplement, vitamin, herbal remedy, or device sold directly to consumers has been proven in rigorous clinical trials to prevent or treat dementia. The FDA has taken action against numerous companies making unsubstantiated claims in this space. If something sounds like a miracle cure, it isn't.

Be skeptical of websites that use the language of science without the substance of it. Words like quantum, cellular, and neuro are often used in marketing to make products sound medical when they are not. Look for sources that cite actual peer-reviewed studies, name the specific research being referenced, and don't have a product to sell you.

Be skeptical of social media as a primary source of medical information. Patient communities on social media can offer invaluable emotional support and practical caregiving wisdom. *They are not reliable sources for treatment decisions.* What worked for one person's parent may not be appropriate, safe, or effective for your family member.

The Sources You Can Trust

A short list of reliable, regularly updated sources on dementia research and care will serve you better than an hour of searching the internet without direction.

1. The Alzheimer's Association at alz.org is the most comprehensive patient-facing resource in the United States. It publishes the annual Alzheimer's Disease Facts and Figures report, maintains updated information about treatments and clinical trials, runs the 24/7 helpline at 1-800-272-3900, and has a news section that covers research developments with appropriate context.

2. The National Institute on Aging at nia.nih.gov/alzheimers maintains patient-friendly information about Alzheimer's and related dementias, including updates on research and the clinical trials it funds.

3. The Alzheimer's Drug Discovery Foundation at alzdiscovery.org publishes a useful Cognitive Vitality resource with evidence-based ratings of various supplements, lifestyle approaches, and interventions. It's written for general readers and is regularly updated.

4. ClinicalTrials.gov is the official registry of clinical trials in the

United States. Every trial that is enrolling, recently completed, or actively reporting results is listed there. You can search by condition, drug name, location, and eligibility criteria.

5. For peer-reviewed research itself, PubMed at pubmed.ncbi.nlm.nih.gov is the National Library of Medicine's database of published scientific studies. Many articles are available free of charge. The abstracts, which summarize the study and its findings, are almost always free even when the full article requires a subscription.

How to Talk to Your Doctor About What You Have Read

Doctors want their patients to be informed, and most physicians welcome questions from patients who have been reading about their condition. The key is how you bring the information into the appointment.

Approaching a conversation with curiosity rather than certainty works better for everyone. 'I read that there are new blood tests that can detect Alzheimer's before symptoms appear. Would that be relevant for our situation?' will go further than 'I need a blood test for amyloid because I read online that everyone should have one.'

If you read something and aren't sure whether it's credible, tell your doctor that. 'I saw a study about a new drug and I'm not sure if it's reliable, but I wanted to ask you about it.' Your doctor can help you interpret what you've found and put it in clinical context.

Bring things in writing when you can. If you found an article or study that prompted your question, print it or bring it up on your phone. That gives your doctor something concrete to respond to rather than a general impression of what you remember reading.

And keep a running list of questions between appointments. Dementia care involves a lot of information, and it's easy to forget what you wanted to ask once you're in the room. A written list helps you use appointment time efficiently.

Staying Current Without Getting Overwhelmed

Following dementia research can become a full-time job if you let it, and for people who are already carrying the emotional and practical weight of this disease, that kind of information overload is genuinely harmful. A few strategies can help you stay informed without burning out.

1. Designate one or two reliable sources and check them on a schedule that works for you, rather than searching the internet every time a new thought occurs to you. The Alzheimer's Association news page and the National Institute on Aging are good choices. Checking them once a week or once a month gives you meaningful updates without the constant noise of daily searches.

2. Consider signing up for email newsletters from organizations you trust. Both the Alzheimer's Association and the Alzheimer's Drug Discovery Foundation offer newsletters that summarize new developments in plain language and send them directly to your inbox. You get the headlines without the algorithm deciding what you see.

3. When you read something that worries or excites you, give yourself a day before acting on it. Most individual studies don't require an immediate response. Taking a breath, noting the source, and then talking to your doctor at your next scheduled appointment is almost always the right approach.

Your Place in the Future of This Disease

I want to close this book, and this series, with something that I believe deeply from decades in clinical practice.

The people reading these books are not passive observers of the dementia story. They are part of it. The ones who enrolled in clinical trials that produced the drugs now being approved. The ones who shared their stories with researchers, donated to advocacy organizations, called their elected officials, and showed up for support groups and kept going. The caregivers who demanded better resources and pushed back when the system failed them. The people with diagnoses who lived with dignity and purpose and showed everyone around them what was possible.

The future of dementia treatment is going to be built by people like the ones reading these pages. It has always been built by people like you.

Stay curious. Stay connected. Keep asking questions. And know that the investment you make in understanding this disease, whether for yourself or for someone you love, is not wasted. Every informed patient, every engaged caregiver, every person who talks honestly about dementia in their community, makes the road a little shorter for the people who come next.

Finding a Clinical Trial: A Step-by-Step Guide

You may not think of yourself as someone who would join a research study. But clinical trials are not just for scientists or people in big cities with access to research hospitals. *They are for people like you*. And the government maintains a *free, public database that lists every active trial in the country,* including many that are actively looking for participants right now.

The Official Database: ClinicalTrials.gov

ClinicalTrials.gov is a website run by the U.S. National Library of Medicine, which is part of the National Institutes of Health. It is the official, government-maintained registry of clinical studies being conducted in the United States and around the world. Every trial that receives federal funding, or that is studying a drug or device that needs FDA review, is required to be listed there. That means you can trust that what you find on this site is real.

As of 2025, ClinicalTrials.gov lists *hundreds of studies focused on Alzheimer's disease, Lewy body dementia, frontotemporal dementia, vascular dementia, and related conditions.* Some are testing new drugs. Some are studying lifestyle changes, sleep, or diet. Some are looking at new ways to diagnose dementia earlier. There is something for people at many different stages of life and disease.

How to Search the Database

Searching ClinicalTrials.gov is ***free*** and *does not require creating an account.* Here is how to do it:

1: Go to the website. Open a web browser and type clinicaltrials .gov into the address bar. The site works on computers, tablets, and smartphones.

2: Use the search box. On the home page, you will see a search bar. Type in the condition you are looking for, such as "Alzheimer's disease," "Lewy body dementia," "frontotemporal dementia," or simply "dementia."

3: Filter by location. After your results appear, look for the filter options on the left side of the page. You can enter your city, state, or ZIP code to find trials happening near you. You can also filter by distance, such as trials within 50 miles or 100 miles of your home.

4: Filter by status. Look for the "Status" filter and select "*Recruiting*." This will show you only the trials that are actively looking for participants right now. Trials marked "Not yet recruiting" or "Completed" are not currently accepting new participants.

5: *Filter by age*. Many dementia trials look for participants in specific age ranges. You can filter results by your age to narrow down studies that may be right for you.

6: *Read the eligibility criteria*. Click on any trial that interests you. Each listing includes a section called "Eligibility Criteria" that explains exactly who can and cannot join. Read this carefully before reaching out. When I worked as the National Clinical Monitor for the first national protocol for a medication for Alzheimer's, I had to review the charts for any individuals a center wanted to enroll. In our protocol, one exclusion was a family history of sudden cardiac death or alcoholism in the individual. We did have to exclude a number of individuals for the latter issue.

What You Will Find in a Trial Listing

Each listing on ***ClinicalTrials.gov*** contains important information. You do not need a medical background to read it. Here is what to look for:

1. The title and a plain-language summary of what the study is testing

2. The sponsor or organization running the study, such as a university, hospital, or drug company

3. The status of the trial, for example, *Recruiting, Active but Not Recruiting, or Complete*

4. The locations where the trial is taking place and contact names or phone numbers at each site

5. The eligibility criteria, which are the specific requirements a participant must meet to join

6. The estimated completion date and how long participation typically lasts

7. Whether the study involves a drug, a device, a lifestyle change, or an observational approach

How to Apply or Express Interest

You cannot enroll in a clinical trial directly through ClinicalTrials.gov. The website is a registry, not an enrollment portal. But it gives you everything you need to take the next step.

Contact the study site directly. Each trial listing includes a "Contacts and Locations" section. Look for the name and phone number of a study coordinator at a location near you. This is the person whose job it is to help potential participants find out if they are eligible. You can call or email them, and they will walk you through the screening process.

Talk to your doctor first. Your neurologist or primary care doctor may already know about trials that are right for your situation. Bring the trial listing you found to your next appointment and ask your doctor what they think. Some trials require a referral from a treating physician, and your doctor can also help you understand the eligibility criteria in plain terms.

Ask the Alzheimer's Association for help. The Alzheimer's Association runs a free service called TrialMatch at clinicaltrials.alz.org. You can enter information about yourself or your loved one and receive a personalized list of trials that may be a good match. Their helpline at 1-800-272-3900 is also available around the clock and can connect you with a staff member who specializes in clinical trial referrals.

What Happens After You Contact a Trial

When you contact a study site, you will typically go through a pre-screening process. This usually starts with a phone call or questionnaire to see if you meet the basic criteria before anyone asks you to travel anywhere. If you pass the initial screen, you may be invited to come in for a more detailed evaluation, which can include medical records review, cognitive testing, blood draws, or imaging.

All of this happens before you commit to anything. Joining a clinical trial is voluntary, and you have the right to ask questions, take time to think, discuss the trial with your family, and say no at any point. A document called the informed consent form explains everything about the study in plain language, including any risks. You should read it carefully and ask the study team to explain anything you do not understand before you sign.

Compensation and Reimbursement

Here is something that surprises many people: *participants in clinical trials are often paid for their time and effort.* This is not a secret, and it is not something to be suspicious of. It is a standard and ethical practice in research, and the payment amounts are reviewed and approved by an independent ethics board before the trial begins.

Compensation varies by study. Some trials pay a small amount for each visit. Others provide a total payment at the end of participation, which can range from a few hundred dollars to several hundred dollars or more for studies that require many visits over a long period. Many trials also reimburse participants for travel, parking, and meals.

The specific compensation for any trial will be listed in the informed consent form and is *something you can ask about directly* when you call the study coordinator. There is nothing wrong with asking. The lead researcher in every clinical trial is given financial support for every patient they enroll at their center. It is not unusual for patients to be given a stipend to attend, which enables them to pay for transportation, gas, or a meal.

Beyond the financial piece, many participants in dementia research report that being part of a study gave them a sense of purpose and connection during a difficult time. You are not just a subject in a study. You are a partner in the search for answers that could help everyone who comes after you.

A Few Things to Watch For

Not every opportunity that calls itself a "clinical trial" is the same. A few points to keep in mind:

1. Legitimate trials are registered on ClinicalTrials.gov. If you hear about a study that does not appear in the registry, ask why before you proceed.

2. You should never have to pay to participate in a clinical trial. If anyone asks you for money to join, that is a red flag.

3. Legitimate trials will always give you time to review the informed consent form before asking you to sign. If anyone pressures you to sign quickly, slow down and ask questions.

4. Your regular medical care cannot be withheld from you as a condition of trial participation. Joining a study should add to your care, not replace it.

Quick Reference: Where to Start

1. ClinicalTrials.gov—the official government registry; search

and filter by condition, location, and status

2. clinicaltrials.alz.org—the Alzheimer's Association TrialMatch tool; personalized matching for dementia trials

3. 1-800-272-3900—the Alzheimer's Association helpline, available 24 hours a day, seven days a week

4. Your neurologist or memory care specialist—always a good first conversation before contacting a trial site

The research that will change the course of this disease is happening right now. And it cannot happen without people who are willing to participate. If a clinical trial is a possibility for you or someone you love, it is worth taking the time to look.

Thank you for being here.

Chapter Summary

Medical news about dementia is frequently misleading because individual studies are reported as if they settle questions permanently, when real scientific progress comes from accumulated evidence over time. Reliable sources for dementia research include the Alzheimer's Association, the National Institute on Aging, the Alzheimer's Drug Discovery Foundation, ClinicalTrials.gov, and PubMed. Patients can bring research questions to their doctors effectively by framing them with curiosity and bringing specific sources to discuss. Staying informed without becoming overwhelmed requires choosing a small number of trusted sources, checking them on a schedule rather than constantly searching, and waiting before acting on alarming or exciting individual headlines. People affected by dementia are active participants in the future of this disease, not passive observers of it.

Three Things You Can Do Right Now

1. Choose two reliable sources for dementia news and bookmark them today. The Alzheimer's Association news page at alz.org and the Alzheimer's Drug Discovery Foundation's Cognitive Vitality page at alzdiscovery.org are excellent starting points.

2. Sign up for email newsletters from one or two organizations you trust so that updates come to you, instead of requiring you to search. The Alzheimer's Association offers a free newsletter with plain-language research updates.

3. Before your next doctor's appointment, write down one question about something you've read or heard about dementia research. Bring it with you. Starting the habit of bringing informed questions to your appointments makes every visit more useful.

Addendum: A Nobel Prize That Could Change Everything

A Puzzle at the Heart of Life

Every cell in your body is packed with proteins, and *each one has to fold into exactly the right shape to do its job*. Think of it like origami. If even one crease goes in the wrong direction, the whole shape collapses. When proteins fold the wrong way inside the brain, they can *clump together and become toxic*. In Alzheimer's disease, two proteins, *amyloid and tau*, do exactly that. They misfold, stick together, and slowly destroy brain cells.

For decades, scientists had no reliable way to predict what shape a protein would take. That single gap in knowledge blocked the path to treatments for diseases like Alzheimer's, Parkinson's, and Lewy body dementia. Researchers could identify the problem, but they couldn't see it clearly enough to create a solution.

That changed in October 2024, when the Nobel Prize in Chemistry was awarded to three scientists whose work, taken together, solved one of biology's oldest mysteries.

Three Scientists, One Landmark Prize

The 2024 Nobel Prize in Chemistry was shared by Demis Hassabis and John Jumper of Google DeepMind, and David Baker of the University of Washington. Their paths to the prize were different, but their achievements point in the same direction: toward a future where the diseases that have defied treatment for generations may finally meet their match.

Hassabis and Jumper developed AlphaFold, *an artificial intelligence system* that can predict the three-dimensional shape of a protein from its chemical building blocks alone. Before AlphaFold, solving the structure of a single protein *could take years* of painstaking laboratory work. ***AlphaFold can do it in minutes***, with extraordinary accuracy.

David Baker contributed something equally remarkable. He figured out how to design entirely new proteins, proteins that don't exist anywhere in nature. Using a program called Rosetta, and later more advanced AI tools, his team has created thousands of custom proteins. These include *molecules that can neutralize viruses, target cancer cells*, and potentially act as precision tools inside the brain.

As Baker said in describing the implications of this work: "*By designing proteins not found in nature, I believe we will solve many urgent challenges across medicine, technology, and sustainability.*"

Why This Matters for Alzheimer's and Related Dementias

Alzheimer's disease is, at its core, a protein problem. Amyloid plaques build up between brain cells. Tau tangles form inside them. Together, they disrupt communication, trigger inflammation, and *cause neurons to die.* Parkinson's disease and Lewy body dementia involve a third misfolded protein called alpha-synuclein, which spreads through the brain and nerve cells in its own destructive pattern.

AlphaFold has now *mapped the shapes of more than 200 million proteins.* More than three million researchers in 190 countries have used the freely available AlphaFold database, and over 30 percent of the research built on it is focused directly on understanding disease. Scientists studying Alzheimer's, Parkinson's, and Lewy body dementia now have a detailed picture of how these damaging proteins are structured, and that picture is the starting point for designing drugs that can interfere with them.

The logic is straightforward: if you know a protein's exact shape, you can design a drug that fits against it the way a key fits a lock. You can look for weak points in its structure. You can block the places where it bonds with other proteins to form toxic clumps.

AlphaFold 3, released in 2024 and now open to researchers worldwide, goes even further. It can model not just individual proteins but how proteins interact with DNA, RNA, and the small molecules that pharmaceutical researchers use as drug candidates. For families living with dementia, that is not a technical detail. *It is a shortcut that could shave years off the search for the next treatment.*

From the Laboratory to Real People

This research is already moving from the laboratory toward real people. At the National Institutes of Health, an AI-driven initiative launched in 2024 identified three compounds originally developed for cardiovascular and metabolic conditions that now show early promise in neurodegenerative disease models. One of them, when tested in a preclinical model of Alzheimer's, *reduced amyloid plaque buildup by 42 percent* over a 12-week period.

Separately, NIH-funded researchers are advancing a small molecule drug called CT1812 through clinical trials. CT1812 appears to work by displacing toxic protein aggregates at the synapses, the spaces between brain cells where communication happens. Importantly, it may act on both beta-amyloid and alpha-synuclein, which means it could have relevance for Alzheimer's and Lewy body dementia at the same time.

David Baker's lab has gone further still. In April 2026, Baker described a technology currently in validation testing that uses *AI-designed nanomachines* to selectively remove tau proteins from brain tissue. He called the selective removal of dementia-causing substances possible. That isn't a headline from a science fiction novel. That is a Nobel laureate describing work happening right now.

What It Means to Have a Map

Perhaps the clearest way to understand why AlphaFold matters is to think about what it was like before it existed. Researchers knew Alzheimer's involved misfolded proteins. They knew those proteins destroyed brain cells. But they were working largely in the dark, unable to see the precise architecture of the enemy they were fighting.

One neuroscientist put it plainly: *what used to take decades or even a lifetime of work can now take a minute*. That isn't an exaggeration. That is what structural prediction at this scale actually means.

For the millions of families living with Alzheimer's and related conditions, AlphaFold and the work it has inspired represent a new kind of hope. Not the cautious, one-step-at-a-time kind of hope that has marked this field for the past several decades. This is something more. It is the difference between searching for a lost key in a dark room and turning on the lights.

The Race Is On

The scientific community is now widely referring to 2026 as the year of tau because so many trials targeting tau protein are reaching readout points at the same time. Clinical research teams at major pharmaceutical companies and research universities are all drawing on the same AI-assisted understanding of protein structure to design their next-generation treatments.

This is not a cure. Not yet. **But it is a moment unlike any other in the history of dementia research**. The tools that for generations were not available are available now. The questions that could not be answered before are being answered. And the families that have watched someone they love fade away, wondering if anything would ever truly change, have something real to point to.

The Nobel Prize in Chemistry 2024 belongs to three scientists. *But the hope it carries belongs to everyone.*

The Map Gets Bigger: One Billion Proteins and Counting

Earlier in this chapter, you learned that AlphaFold had mapped more than 200 million proteins, a number that once seemed almost impossible to take in. At the time, scientists called it one of the greatest leaps in biology in decades. Now, just a short time later, that number already looks modest.

In May 2026, a team of researchers at an institute called Biohub, based in San Francisco and backed by the Chan Zuckerberg Initiative, announced something that stopped the scientific world in its tracks. Using a new artificial intelligence tool called ESMFold2, they had predicted the shapes of more than one billion proteins. **That is one billion**, with a "B." The same team also mapped information on the sequences of nearly *seven billion additional proteins*, sequences that had been sitting in databases for years but had never been closely studied or understood.

A Map of the Whole Planet

Here is a way to picture what this means. AlphaFold gave science a detailed, reliable map of a continent. ESMFold2 has now handed researchers a map of the entire planet, plus detailed notes on territory that no one had ever explored before.

Many of the newly mapped proteins came from what scientists call metagenomic sources. That is a technical way of saying they were collected from places like soil, ocean water, and other natural environments. These are proteins from living organisms that had been gathered but never fully characterized. By predicting their shapes, ESMFold2 opens entirely new ground for scientific exploration.

What This Means for Alzheimer's Research

This is not simply a numbers competition. More proteins mapped means more chances to identify a protein that plays a role in Alzheimer's disease, to see how it behaves, and to design a treatment that targets it precisely. Researchers studying Alzheimer's, Parkinson's, and other brain diseases now have a far larger library of protein shapes to search through, compare, and work with. The bigger the library, the better the chances of finding a match.

ESMFold2 has also shown it can outperform AlphaFold3, the most recent version of Google DeepMind's system, at one particularly important task: *predicting how proteins interact with each other.* That matters greatly in Alzheimer's research, because the disease involves proteins that clump together and cause damage. Understanding exactly how that clumping happens at the molecular level is a critical step toward stopping it.

The tool has already been used to design new antibodies, which are proteins the body uses to fight threats, that can attach tightly to proteins linked to cancer and immune conditions. When those designs were tested in the laboratory, a high percentage of them performed exactly as predicted. That kind of precision is precisely what drug developers working on Alzheimer's treatments need.

Free and Open to Everyone

One more detail makes this advance stand out. ESMFold2 is fully open source, meaning *any scientist anywhere in the world can use it at no cost.* That matters because some of the most promising research on Alzheimer's and related conditions takes place at universities and smaller institutions that do not have the budgets of large pharmaceutical companies. Removing cost and access barriers means more minds working on the same problem, and that tends to speed things up.

The Pace of Change

The speed of this progress is worth pausing to consider. AlphaFold earned a Nobel Prize for mapping 200 million proteins. ESMFold2, building on that foundation, has now surpassed that number by more than 800 million entries, and it arrived just a few years later. The technology is not standing still. It is moving faster than most researchers predicted even recently.

For families living with Alzheimer's disease, that pace is meaningful. Every advance like this one brings the possibility of effective treatment a little closer. The map of the protein universe is getting bigger by the year. And the bigger the map, the better the chances that researchers will find exactly what they need.

References

Abramson, J., Adler, J., Dunger, J., Evans, R., Green, T., Pritzel, A., Ronneberger, O., Willmore, L., Ballard, A. J., Bambrick, J., Bodenstein, S. W., Evans, D. A., Hung, C.-C., O'Neill, M., Reiman, D., Tunyasuvunakool, K., Wu, Z., Zemgulyte, A., Arvaniti, E., & Jumper, J. M. (2024). Accurate structure prediction of biomolecular interactions with AlphaFold 3. Nature, 630, 493–500. https://doi.org/10.1038/s41586-024-07487-w

Alzheimer, A. (1907). Uber eine eigenartige Erkrankung der Hirnrinde [About a peculiar disease of the cerebral cortex]. Allgemeine Zeitschrift fur Psychiatrie und Psychisch-Gerichtliche Medizin, 64, 146–148.

Alzheimer's Association. (2023). Stages of Alzheimer's disease. https://www.alz.org/alzheimers-dementia/stages

Alzheimer's Association. (2024). 2024 Alzheimer's disease facts and figures. Alzheimer's & Dementia, 20(5), 3708–3821. https://doi.org/10.1002/alz.13809

Alzheimer's Association. (2025). 2025 Alzheimer's disease facts and figures. Alzheimer's & Dementia, 21, e70235. https://doi.org/10.1002/alz.70235

Alzheimer's Association. (2024). TrialMatch: Find an Alzheimer's clinical trial. https://trialmatch.alz.org

Alzheimer's Disease International. (2023). World Alzheimer report 2023: Reducing dementia risk: Never too early, never too late. Alzheimer's Disease International.

Bherer, L., Erickson, K. I., & Liu-Ambrose, T. (2013). A review of the effects of physical activity and exercise on cognitive and brain functions in older adults. Journal of Aging Research, 2013, Article 657508. https://doi.org/10.1155/2013/657508

BioSpace. (2026, January 19). 4 next-gen candidates that could form the future of Alzheimer's treatment. https://www.biospace.com/drug-development/4-next-gen-candidates-that-could-form-the-future-of-alzheimers-treatment

Blennow, K., de Leon, M. J., & Zetterberg, H. (2006). Alzheimer's disease. The Lancet, 368(9533), 387–403. https://doi.org/10.1016/S0140-6736(06)69113-7

Braak, H., & Braak, E. (1991). Neuropathological stageing of Alzheimer-related changes. Acta Neuropathologica, 82(4), 239–259. https://doi.org/10.1007/BF00308809

Callaway, E., & Naddaf, M. (2026, May 27). Move over, AlphaFold: Open source model predicts shape of 1 billion proteins. Nature. https://doi.org/10.1038/d41586-026-01686-3

Centers for Disease Control and Prevention. (2023). Alzheimer's disease and healthy aging. U.S. Department of Health and Human Services. https://www.cdc.gov/aging/aginginfo/alzheimers.htm

Cheung, C. Y., Ran, A. R., Wang, S., Chan, V. T., Sham, K., Hilal, S., Venketasubramanian, N., Cheng, C. Y., Sabanayagam, C., Tham, Y. C., Schmetterer, L., Ikram, M. K., Chen, C. L., & Wong, T. Y. (2022). A deep learning model for detection of Alzheimer's disease based on retinal photographs: A retrospective, multicentre case-control study. The Lancet Digital Health, 4(11), e806–e815. https://doi.org/10.1016/S2589-7500(22)00169-8

Corder, E. H., Saunders, A. M., Strittmatter, W. J., Schmechel, D. E., Gaskell, P. C., Small, G. W., Roses, A. D., Haines, J. L., & Pericak-Vance, M. A. (1993). Gene dose of apolipoprotein E type 4 allele and the risk of Alzheimer's disease in late onset families. Science, 261(5123), 921–923. https://doi.org/10.1126/science.8346443

Cryan, J. F., O'Riordan, K. J., Cowan, C. S. M., Sandhu, K. V., Bastiaanssen, T. F. S., Boehme, M., Codagnone, M. G., Cussotto, S., Fulling, C., Golubeva, A. V., Guzzetta, K. E., Jaggar, M., Long-Smith, C. M., Lyte, J. M., Martin, J. A., Molinero-Perez, A., Moloney, G., Morelli, E., Morillas, E., & Dinan, T. G. (2019). The microbiota-gut-brain axis. Physiological Reviews, 99(4), 1877–2013. https://doi.org/10.1152/physrev.00018.2018

Cummings, J., Lee, G., Zhong, K., Fonseca, J., & Taghva, K. (2021). Alzheimer's disease drug development pipeline: 2021. Alzheimer's & Dementia: Translational Research & Clinical Interventions, 7(1), Article e12179. https://doi.org/10.1002/trc2.12179

DeVos, S. L., Miller, R. L., Schoch, K. M., Holmes, B. B., Kebodeaux, C. S., Wegmann, S., Hoppe, B., Chu, D., Mahan, T. E., Bhattacharya, S., & Miller, T. M. (2017). Tau reduction prevents neuronal loss and reverses pathological tau deposition and seeding in mice with tauopathy. Science Translational Medicine, 9(374), Article eaag0481. https://doi.org/10.1126/scitranslmed.aag0481

Dubois, B., Villain, N., Schneider, L., Fox, N., Campbell, N., Galasko, D., Kivipelto, M., Jessen, F., Hanseeuw, B., Boada, M., Barkhof, F., Nordberg, A., Froelich, L., Waldemar, G., Frederiksen, K. S., Padovani, A., Planche, V., Rowe, C., Bejanin, A., & Frisoni, G. B. (2024). Alzheimer disease as a clinical-biological construct: An

International Working Group recommendation. JAMA Neurology, 81(12), 1304–1311. https://doi.org/10.1001/jamaneurol.2024.3770

Erickson, K. I., Voss, M. W., Prakash, R. S., Basak, C., Szabo, A., Chaddock, L., Kim, J. S., Heo, S., Alves, H., White, S. M., Wojcicki, T. R., Mailey, E., Vieira, V. J., Martin, S. A., Pence, B. D., Woods, J. A., McAuley, E., & Kramer, A. F. (2011). Exercise training increases size of hippocampus and improves memory. Proceedings of the National Academy of Sciences, 108(7), 3017–3022. https://doi.org/10.1073/pnas.1015950108

Fraser, K. C., Meltzer, J. A., & Rudzicz, F. (2016). Linguistic features identify Alzheimer's disease in narrative speech. Journal of Alzheimer's Disease, 49(2), 407–422. https://doi.org/10.3233/JAD-150520

Fratiglioni, L., Paillard-Borg, S., & Winblad, B. (2004). An active and socially integrated lifestyle in late life might protect against dementia. The Lancet Neurology, 3(6), 343–353. https://doi.org/10.1016/S1474-4422(04)00767-7

Google DeepMind. (2026, March 3). AlphaFold: Five years of impact. https://deepmind.google/blog/alphafold-five-years-of-impact/

Gorelick, P. B., Scuteri, A., Black, S. E., DeCarli, C., Greenberg, S. M., Iadecola, C., Launer, L. J., Laurent, S., Lopez, O. L., Nyenhuis, D., Petersen, R. C., Schneider, J. A., Tzourio, C., Arnett, D. K., Bennett, D. A., Chui, H. C., Higashida, R. T., Lindquist, R., Nilsson, P. M., & Seshadri, S. (2011). Vascular contributions to cognitive impairment and dementia: A statement for healthcare professionals from the American Heart Association/American Stroke Association. Stroke, 42(9), 2672–2713. https://doi.org/10.1161/STR.0b013e3182299496

Guerreiro, R., Wojtas, A., Bras, J., Carrasquillo, M., Rogaeva, E., Majounie, E., Cruchaga, C., Sassi, C., Kauwe, J. S., Younkin, S.,

Hazrati, L., Collinge, J., Pocock, J., Lashley, T., Williams, J., Lambert, J. C., Amouyel, P., Goate, A., Rademakers, R., & Hardy, J. (2013). TREM2 variants in Alzheimer's disease. New England Journal of Medicine, 368(2), 117–127. https://doi.org/10.1056/NEJMoa1211851

Hansson, O. (2021). Biomarkers for neurodegenerative diseases. Nature Medicine, 27(6), 954–963. https://doi.org/10.1038/s41591-021-01382-x

Hardy, J. A., & Higgins, G. A. (1992). Alzheimer's disease: The amyloid cascade hypothesis. Science, 255(5043), 184–186. https://doi.org/10.1126/science.1566067

Iliff, J. J., Wang, M., Liao, Y., Plogg, B. A., Peng, W., Gundersen, G. A., Benveniste, H., Vates, G. E., Deane, R., Goldman, S. A., Nagelhus, E. A., & Nedergaard, M. (2012). A paravascular pathway facilitates CSF flow through the brain parenchyma and the clearance of interstitial solutes, including amyloid β. Science Translational Medicine, 4(147), Article 147ra111. https://doi.org/10.1126/scitranslmed.3003748

Jack, C. R., Jr., Bennett, D. A., Blennow, K., Carrillo, M. C., Dunn, B., Haeberlein, S. B., Holtzman, D. M., Jagust, W., Jessen, F., Karlawish, J., Liu, E., Molinuevo, J. L., Montine, T., Phelps, C., Rankin, K. P., Rowe, C. C., Siemers, E., Snyder, H. M., Sperling, R., & Contributors. (2018). NIA-AA Research Framework: Toward a biological definition of Alzheimer's disease. Alzheimer's & Dementia, 14(4), 535–562. https://doi.org/10.1016/j.jalz.2018.02.018

Jumper, J., Evans, R., Pritzel, A., Green, T., Figurnov, M., Ronneberger, O., Tunyasuvunakool, K., Bates, R., Zidek, A., Potapenko, A., Bridgland, A., Meyer, C., Kohl, S. A. A., Ballard, A. J., Cowie, A., Romera-Paredes, B., Nikolov, S., Jain, R., Adler, J., & Hassabis, D.

(2021). Highly accurate protein structure prediction with AlphaFold. Nature, 596, 583–589. https://doi.org/10.1038/s41586-021-03819-2

Komor, A. C., Badran, A. H., & Liu, D. R. (2017). CRISPR-based technologies for the manipulation of eukaryotic genomes. Cell, 168(1–2), 20–36. https://doi.org/10.1016/j.cell.2016.10.044

Kourtis, L. C., Regele, O. B., Wright, J. M., & Jones, G. B. (2019). Digital biomarkers for Alzheimer's disease: The mobile/wearable devices opportunity. NPJ Digital Medicine, 2, Article 9. https://doi.org/10.1038/s41746-019-0084-2

Lam, K., Lu, A. D., Shi, Y., & Covinsky, K. E. (2020). Assessing telemedicine unreadiness among older adults in the United States during the COVID-19 pandemic. JAMA Internal Medicine, 180(10), 1389–1391. https://doi.org/10.1001/jamainternmed.2020.2671

Lin, F. R., Yaffe, K., Xia, J., Xue, Q. L., Harris, T. B., Purchase-Helzner, E., Satterfield, S., Ayonayon, H. N., Ferrucci, L., & Simonsick, E. M. (2013). Hearing loss and cognitive decline in older adults. JAMA Internal Medicine, 173(4), 293–299. https://doi.org/10.1001/jamainternmed.2013.1868

Lin, F. R., Pike, J. R., Albert, M. S., Arnold, M., Burgard, S., Chisolm, T., Couper, D., Deal, J. A., Goman, A. M., Glynn, N. W., Gmelin, T., Gravens-Mueller, L., Hayden, K. M., Huang, A. R., Knopman, D., Mitchell, C. M., Mosley, T., Pankow, J. S., Reed, N. S., Sanchez, V., Schrack, J. A., Windham, B. G., & Coresh, J. (2023). Hearing intervention versus health education control to reduce cognitive decline in older adults with hearing loss in the USA (ACHIEVE): A multicentre, randomised controlled trial. The Lancet, 402(10404), 786–797. https://doi.org/10.1016/S0140-6736(23)01406-X

Livingston, G., Sommerlad, A., Orgeta, V., Costafreda, S. G., Huntley, J., Ames, D., Ballard, C., Banerjee, S., Burns, A., Cohen-Mansfield, J., Cooper, C., Fox, N., Gitlin, L. N., Howard, R., Kales, H. C., Larson, E. B., Ritchie, K., Rockwood, K., Sampson, E. L., & Mukadam, N. (2017). Dementia prevention, intervention, and care. The Lancet, 390(10113), 2673–2734. https://doi.org/10.1016/S0140-6736(17)31363-6

Livingston, G., Huntley, J., Sommerlad, A., Ames, D., Ballard, C., Banerjee, S., Brayne, C., Burns, A., Cohen-Mansfield, J., Cooper, C., Costafreda, S. G., Dias, A., Fox, N., Gitlin, L. N., Howard, R., Kales, H. C., Kivimaki, M., Larson, E. B., Ogunniyi, A., & Mukadam, N. (2020). Dementia prevention, intervention, and care: 2020 report of the Lancet Commission. The Lancet, 396(10248), 413–446. https://doi.org/10.1016/S0140-6736(20)30367-6

Livingston, G., Huntley, J., Liu, K. Y., Costafreda, S. G., Selbaek, G., Alladi, S., Ames, D., Banerjee, S., Burns, A., Brayne, C., Fox, N. C., Ferri, C. P., Gitlin, L. N., Howard, R., Kales, H. C., Kivimaki, M., Larson, E. B., Nakasujja, N., Rockwood, K., & Mukadam, N. (2024). Dementia prevention, intervention, and care: 2024 report of the Lancet standing Commission. The Lancet, 404(10452), 572–628. https://doi.org/10.1016/S0140-6736(24)01296-0

McKeith, I. G., Boeve, B. F., Dickson, D. W., Halliday, G., Taylor, J. P., Weintraub, D., Aarsland, D., Galvin, J., Attems, J., Ballard, C. G., Bayston, A., Beach, T. G., Blanc, F., Bohnen, N., Bonanni, L., Bras, J., Brundin, P., Burn, D., Chen-Plotkin, A., & Kosaka, K. (2017). Diagnosis and management of dementia with Lewy bodies: Fourth consensus report of the DLB Consortium. Neurology, 89(1), 88–100. https://doi.org/10.1212/WNL.0000000000004058

Morris, M. C., Tangney, C. C., Wang, Y., Sacks, F. M., Bennett, D. A., & Aggarwal, N. T. (2015). MIND diet associated with reduced incidence of Alzheimer's disease. Alzheimer's & Dementia, 11(9), 1007–1014. https://doi.org/10.1016/j.jalz.2014.11.009

National Institute on Aging. (2023). Alzheimer's disease fact sheet. U.S. Department of Health and Human Services. https://www.nia.nih.gov/health/alzheimers-disease-fact-sheet

National Institute on Aging. (2025). 2025 NIH Alzheimer's disease and related dementias research progress report: Advances and achievements. U.S. Department of Health and Human Services. https://www.nia.nih.gov/about/2025-nih-dementia-research-progress-report

Ngandu, T., Lehtisalo, J., Solomon, A., Levalahti, E., Ahtiluoto, S., Antikainen, R., Backman, L., Hanninen, T., Jula, A., Laatikainen, T., Lindstrom, J., Mangialasche, F., Paajanen, T., Pajala, S., Peltonen, M., Rauramaa, R., Stigsdotter-Neely, A., Strandberg, T., Tuomilehto, J., & Kivipelto, M. (2015). A 2 year multidomain intervention of diet, exercise, cognitive training, and vascular risk monitoring versus control to prevent cognitive decline in at-risk elderly people (FINGER): A randomised controlled trial. The Lancet, 385(9984), 2255–2263. https://doi.org/10.1016/S0140-6736(15)60461-5

Palmqvist, S., Janelidze, S., Quiroz, Y. T., Zetterberg, H., Lopera, F., Stomrud, E., Su, Y., Chen, Y., Serrano, G. E., Leuzy, A., Mattsson-Carlgren, N., Strandberg, O., Smith, R., Villegas, A., Sepulveda-Falla, D., Chai, X., Proctor, N. K., Beach, T. G., Blennow, K., & Hansson, O. (2020). Discriminative accuracy of plasma phospho-tau217 for Alzheimer disease vs other neurodegenerative disorders. JAMA, 324(8), 772–781. https://doi.org/10.1001/jama.2020.12134

Petersen, R. C. (2004). Mild cognitive impairment as a diagnostic entity. Journal of Internal Medicine, 256(3), 183–194. https://doi.org/10.1111/j.1365-2796.2004.01388.x

Rascovsky, K., Hodges, J. R., Knopman, D., Mendez, M. F., Kramer, J. H., Neuhaus, J., van Zandijcke, M., Stieltjes, B., Grossman,

M., Miller, B. L., & Levenson, R. L. (2011). Sensitivity of revised diagnostic criteria for the behavioural variant of frontotemporal dementia. Brain, 134(9), 2456–2477. https://doi.org/10.1093/brain/awr179

The Royal Swedish Academy of Sciences. (2024, October 9). The Nobel Prize in Chemistry 2024 [Press release]. NobelPrize.org. https://www.nobelprize.org/prizes/chemistry/2024/press-release/

Selkoe, D. J., & Hardy, J. (2016). The amyloid hypothesis of Alzheimer's disease at 25 years. EMBO Molecular Medicine, 8(6), 595–608. https://doi.org/10.15252/emmm.201606210

Seoul Economic Daily. (2026, April 6). Nobel Laureate Baker: "AI can create proteins not found in nature." https://en.sedaily.com/news/2026/04/06/nobel-laureate-baker-ai-can-create-proteins-not-found-in

Sims, J. R., Zimmer, J. A., Evans, C. D., Lu, M., Ardayfio, P., Sparks, J., Wessels, A. M., Shcherbinin, S., Wang, H., Monkul Nery, E. S., Collins, E. C., Solomon, P., Salloway, S., Apostolova, L. G., Hansson, O., Ritchie, C., Brooks, D. A., Mintun, M., Skovronsky, D. M., & the TRAILBLAZER-ALZ 2 Investigators. (2023). Donanemab in early symptomatic Alzheimer's disease: The TRAILBLAZER-ALZ 2 randomized clinical trial. JAMA, 330(6), 512–527. https://doi.org/10.1001/jama.2023.13239

Sperling, R. A., Aisen, P. S., Beckett, L. A., Bennett, D. A., Craft, S., Fagan, A. M., Iwatsubo, T., Jack, C. R., Jr., Kaye, J., Montine, T. J., Park, D. C., Reiman, E. M., Rowe, C. C., Siemers, E., Stern, Y., Yaffe, K., Carrillo, M. C., Thies, B., Morrison-Bogorad, M., & Phelps, C. H. (2011). Toward defining the preclinical stages of Alzheimer's disease: Recommendations from the National Institute on Aging-Alzheimer's Association workgroups on diagnostic guidelines for Alzheimer's disease. Alzheimer's & Dementia, 7(3), 280–292. https://doi.org/10.1016/j.jalz.2011.03.003

Stern, Y. (2012). Cognitive reserve in ageing and Alzheimer's disease. The Lancet Neurology, 11(11), 1006–1012. https://doi.org/10.1016/S1474-4422(12)70191-6

U.S. National Library of Medicine. (2024). ClinicalTrials.gov. https://www.clinicaltrials.gov

University of Washington Institute for Protein Design. (2024, October 9). David Baker wins Nobel Prize for protein design. https://www.ipd.uw.edu/2024/10/david-baker-wins-nobel-prize-for-protein-design/

van Dyck, C. H., Swanson, C. J., Aisen, P., Bateman, R. J., Chen, C., Gee, M., Kanekiyo, M., Li, D., Reyderman, L., Cohen, S., Froelich, L., Katayama, S., Sabbagh, M., Vellas, B., Watson, D., Dhadda, S., Irizarry, M., Kramer, L. D., & Iwatsubo, T. (2023). Lecanemab in early Alzheimer's disease. New England Journal of Medicine, 388(1), 9–21. https://doi.org/10.1056/NEJMoa2212948

Wada, K., & Shibata, T. (2007). Living with seal robots: Its sociopsychological and physiological influences on the elderly at a care house. IEEE Transactions on Robotics, 23(5), 972–980. https://doi.org/10.1109/TRO.2007.906261

Wang, L., Davis, P. B., Volkow, N. D., Berger, N. A., Kaelber, D. C., & Xu, R. (2024). Associations of semaglutide with first-time diagnosis of Alzheimer's disease in patients with type 2 diabetes: Target trial emulation using nationwide real-world data in the United States. Alzheimer's & Dementia, 20(8), 4745–4754. https://doi.org/10.1002/alz.14313

Williamson, J. D., Pajewski, N. M., Auchus, A. P., Bryan, R. N., Chelune, G., Cheung, A. K., Cleveland, M. L., Coker, L. H., Crowe, M. G., Cushman, W. C., Cutler, J. A., Davatzikos, C., Desiderio, L., Ellenberg, J. H., Fine, L. J., Gaussoin, S. A., Harris, D., Harwell, L., & Murray, A. M. (2019). Effect of intensive vs standard blood pressure control on probable dementia: A randomized clinical trial. JAMA, 321(6), 553–561. https://doi.org/10.1001/jama.2018.21442

Xie, L., Kang, H., Xu, Q., Chen, M. J., Liao, Y., Thiyagarajan, M., O'Donnell, J., Christensen, D. J., Nicholson, C., Iliff, J. J., Takano, T., Deane, R., & Nedergaard, M. (2013). Sleep drives metabolite clearance from the adult brain. Science, 342(6156), 373–377. https://doi.org/10.1126/science.1241224

Yaffe, K., Barnes, D., Nevitt, M., Lui, L. Y., & Covinsky, K. (2001). A prospective study of physical activity and cognitive decline in elderly women. Archives of Internal Medicine, 161(14), 1703–1708. https://doi.org/10.1001/archinte.161.14.1703

Zetterberg, H., & Blennow, K. (2021). Moving fluid biomarkers for Alzheimer's disease from research tools to routine clinical diagnostics. Molecular Neurodegeneration, 16, Article 10. https://doi.org/10.1186/s13024-021-00430-x

About the Author

Dr. Patricia A. Farrell is a licensed psychologist, published author of multiple self-help books and videos, former WebMD psychologist expert/consultant, former medical consultant for Social Security Disability Determinations, psychiatric researcher at Mt. Sinai Medical Center (NYC) and an educator who has taught on the college as well as graduate and post- graduate levels, and is a top health writer for Medium.com publications.

Dr. Farrell is a consultant to pharmaceutical firms, writes continuing education modules for mental healthcare professionals, has contributed to USMLE medical school prep courses and blogs on M edium.com and on Patreon "Dr. Farrell Unplugged," has a YouTube channel with helpful tips on multiple topics and is a biographee in Who's Who in the World, Who's Who in America and Who's Who in American Women.

A member of The American Psychological Association as well as the SAG-AFTRA union, Dr. Farrell is a former board member of the NJ Board of Psychological Examiners, a former psychiatry preceptor at UMDNJ and a former board of directors' member of Bergen Pines Hospital.

A Special Request

If this book has touched your heart, sparked your curiosity, or simply entertained you along the way, I'd be incredibly grateful if you could take a moment to share your thoughts with a review on Amazon or wherever you discovered this book. Your words not only help other readers find books they'll love, but they also mean the world to authors like me who pour their hearts into every page. Thank you for being part of this journey and for helping books find their way to the readers who need them most.

Author Page on Amazon: ***https://tinyurl.com/zbrb6c9n*** and on Draft2Digital: ***https://tinyurl.com/mwta4w46***

www.ingramcontent.com/pod-product-compliance
Lightning Source LLC
LaVergne TN
LVHW030727061126
851857LV00019B/2021
9798994241950